Biology and Mathematics

Series Editor
Marie-Christine Maurel

Biology and Mathematics

History and Challenges

Roger Buis

WILEY

First published 2019 in Great Britain and the United States by ISTE Ltd and John Wiley & Sons, Inc.

Apart from any fair dealing for the purposes of research or private study, or criticism or review, as permitted under the Copyright, Designs and Patents Act 1988, this publication may only be reproduced, stored or transmitted, in any form or by any means, with the prior permission in writing of the publishers, or in the case of reprographic reproduction in accordance with the terms and licenses issued by the CLA. Enquiries concerning reproduction outside these terms should be sent to the publishers at the undermentioned address:

ISTE Ltd
27-37 St George's Road
London SW19 4EU
UK

www.iste.co.uk

John Wiley & Sons, Inc.
111 River Street
Hoboken, NJ 07030
USA

www.wiley.com

© ISTE Ltd 2019

The rights of Roger Buis to be identified as the author of this work have been asserted by him in accordance with the Copyright, Designs and Patents Act 1988.

Library of Congress Control Number: 2019943329

British Library Cataloguing-in-Publication Data
A CIP record for this book is available from the British Library
ISBN 978-1-78630-483-4

Contents

Foreword ix

Introduction xv

Chapter 1. On the Status of Biology: On the Definition of Life 1

    1.1. Causality in biology 4
        1.1.1. Vitalism 8
        1.1.2. Teleology 10
    1.2. Variability in biology 13
        1.2.1. Time-dependence of biological processes 15
        1.2.2. Environment-dependence of biological processes 17

Chapter 2. On the Nature of the Contribution Made by Mathematics to Biology 19

    2.1. On the affinity of mathematics with biology 20
    2.2. Mathematics, an instrument of work and thought on biology 25

Chapter 3. Some Historical Reference Points: Biology Fashioned by Mathematics 35

    3.1. The first remarkable steps in biomathematics 37
        3.1.1. On the continuous in biology 37
        3.1.2. On the discrete in biology 39
        3.1.3. The notion of laws in biology 43
        3.1.4. The beginning of classical science: Descartes and Pascal 44
        3.1.5. Buffon and hesitations relating to the utility of mathematics in biology 45

3.2. Some pertinent contributions from mathematics in the modern era 48
    3.2.1. The laws of growth 48
    3.2.2. Formal genetics 49
3.3. Introduction of the notion of a probabilistic model in biology 56
3.4. The physiology of C. Bernard (1813–1878): the call
to mathematics 58
3.5. The principle of optimality in biology 60
3.6. Introduction of the formalism of dynamic systems in biology 61
3.7. Morphogenesis: the need for mathematics in the study
of biological forms 63
    3.7.1. General principles from D'Arcy Thompson 64
    3.7.2. Turing's reaction–diffusion systems (1952): morphogenesis,
a "break of symmetry" 69
3.8. The theory of automatons and cybernetics in biology 70
    3.8.1. The theory of automatons 70
    3.8.2. The contribution of cybernetics 73
    3.8.3. The case of L-systems 74
    3.8.4. Petri's networks 74
3.9. Molecular biology 78
    3.9.1. On genetic information 81
    3.9.2. The linguistic model in biology 83
3.10. Information and communication, important notions in biology 84
3.11. The property of self-organization in biology 86
    3.11.1. Structural self-organization 87
    3.11.2. Self-reproductive hypercycle 88
3.12. Systemic biology 89
    3.12.1. On the notion of system 89
    3.12.2. Essay in relational biology 90
    3.12.3. Emergence and complexity 93
    3.12.4. Networks 98
    3.12.5. Order, innovation and complex networks 104
3.13. Game theory in biology 105
3.14. Artificial life 109
    3.14.1. Biomimetic automatons 110
    3.14.2. Psychophysiology and mathematics: controls on learning 111
3.15. Bioinformatics 112

Chapter 4. Laws and Models in Biology 115

4.1. Biological laws in literary language 118
    4.1.1. The law of Cuvier's organic correlations (1825) 118
    4.1.2. The fundamental biogenetic law 118

4.2. Biological laws in mathematical language 119
   4.2.1. Statistical laws 121
4.3. Theoretical laws 131
   4.3.1. Formal genetics 131
   4.3.2. Growth laws 132
   4.3.3. Population dynamics 133

Chapter 5. Mathematical Tools and Concepts in Biology 135

5.1. An old biomathematical subject: describing and/or
explaining phyllotaxis 136
5.2. The notion of invariant and its substrate: time and space 140
   5.2.1. Physical time/biological time 142
   5.2.2. Metric space/non-metric space 143
   5.2.3. Multi-scale processes 147
5.3. Continuous formalism 147
   5.3.1. Dynamics of a univariate process 148
   5.3.2. Structured models 149
   5.3.3. Oscillatory dynamics 151
   5.3.4. On the stability of dynamic systems 154
   5.3.5. Multivariate structured models 160
   5.3.6. Dynamics of spatio-temporal process 163
   5.3.7. Multi-scale models 171
5.4. Discreet formalism 174
5.5. Spatialized models 175
   5.5.1. Multi-agent models: dynamics of a biological association
   of the individual-centered type 175
   5.5.2. Electrophysiological models: transmission of electrical signals 176
5.6. Random processes in biology 178
   5.6.1. Poisson process 181
   5.6.2. Birth–death processes 182
5.7. Logic kinetics of regulation 184

Conclusion 189

Glossary 201

References 217

Index 221

Foreword

Roger Buis, professor emeritus at the University of Toulouse (INP – *Institut National Polytechnique*) – already the author of a considerable amount of research into the biomathematics of growth[1] – delivers here a true panorama of the relationships between biology and mathematics over time, and in particular over the course of the last century, accompanied by a series of profound epistemological thoughts, thereby creating a book of great rarity and value.

As we know, mathematics is ancient, just like the interest taken in living things. But the word "biology" only appeared at the beginning of the 19th Century and, whilst E. Kant has already confirmed – and as Roger Buis rightly reminds us – a piece of knowledge is scientific insofar as mathematics has been integrated into it, the explicit idea of applying mathematics to biology is found only with C. Bernard[2] – one of the great references in the book.

In contrast with physics, biology resists mathematization, for understandable reasons: the variability of living things, their dependence on time and on the environment, diversity and the complexity of biological processes, the diffuse aspect of causality (sometimes circular) and the difficulty of mastering the operational conditions of experiments have made obtaining consistencies problematic. Hence,

---

1 Buis, R. (2016). *Biomathématiques de la croissance, le cas des végétaux*. EDP-Sciences, Les Ulis.

2 Let us note that, previously, G.-L. Buffon regretted that qualitative mathematics remained in limbo and could not be applied in natural sciences. "Everything that has an immediate relationship with a position is totally missing from mathematical sciences. The art that Leibniz coined *Analysis situs* has not yet come to light and yet this art that would allow us to discover the relationships of position between things would also be useful and perhaps more necessary for natural sciences than the art whose only objective is the size of things; because we more often need to know the form than the matter." Refer to Buffon, G.-L. (1774). *Œuvres complètes*, vol. IV, chap. IX. Imprimerie Royale, Paris, 73.

the overall appraisal which may be interpreted as disappointing: apart from some specific sectors (in particular genetics), few laws are proven in biology, and even fewer that express themselves in mathematical language.

Roger Buis comments on this, but, going further than the "epistemological obstacles" – coined by G. Bachelard – he takes up the challenge. Even though the description in vernacular language (important in both natural history and Husserlian phenomenology), will always remain the most important in biology, in this discipline we more readily use "modeling" than "demonstration". Nevertheless, mathematics has transferable applications to biology: beyond the savings made by the move to symbols, using mathematical language is not simply using a "language", but a true instrument of thinking, of a remarkable tool of intelligibility which, whilst allowing hypotheses to be clearly laid down, will verify the conclusions by the same amount. Because – let us not doubt it – in biology and elsewhere, the scientific approach is always hypothetico-deductive. Whilst certain preliminary conjectures are less significant here than in physics – for example, the choice of a reference frame (dominant, certainly, in factor analysis, but hardly relevant, in general, elsewhere in biology) – others like approximations or simplifications that we will cautiously allow ourselves to use (linearization, or even quasi-stationarity of certain processes) are essential in this and necessarily lead to significant consequences. At least an advantage is drawn from this: modeling allows controlled experimentation. Thus, the modification of a parameter in a model that is elsewhere structurally stable is going to be possible at will. Mathematics, as a result, does not only provide symbolisms. It also contributes concepts and operating modes that allow real life to be simulated[3].

By examining history, we also realize that, to use the expression of Roger Buis, mathematics has "sculpted" biology. From the point of view of the continuous, geometry, since ancient times, has given rise to the consideration of symmetries and continuous transformations, implicitly presented by Aristotle with the ago-antagonistic couple of power and action. The first separate formalizations appeared from the medieval period onwards, with the famous Fibonacci sequence, which, founded on strong hypotheses, provided the first model of the growth of a population (as it happened, rabbits). Then, in the Classical epoch, there was the era of the first phyllotaxic "laws" relating to the growth speeds of stems or leaves of

---

3 The difference between modeling and simulating can be characterized rapidly as follows: the simulation is not opposed to modeling. Simulations (numerical or computerized) are types of modeling, "model calculations", but they appear most often when there is no shorter method for evaluation of the result than simulating step by step the behavior of the model. To this end, they are "phenomenological models of the behavior of the model", sorts of "models of models", "square" models. Refer to Varenne, F. (2008). Epistemology of models and simulations: Overview and trends. In *Les Modèles: possibilités et limites*, Lévy, J.-M (dir.). Éditions Matériologiques, Paris.

plants, as well as to the mechanics of wood and its constraints (G.-L. Buffon, L. Euler). Finally, during the 19th Century and especially during the 20th Century, calculation of probabilities was based on development of Mendelian genetics, then of the genetics of populations, and subsequently on statistical biometrics (R. Fisher). Whereas C. Bernard highlighted the stationarity of the interior environment of living organisms, principles of optimality, underpinned by the calculation of variations (seeking the extremums of a functionality), are going to become dominant in biology, in particular in plant biology. Then formalisms from system theory (L. von Bertalanffy) encouraged A. J. Lotka and V. Volterra to model the dynamics of interactions between species (prey–predator systems, parasitism) with differential equations. At around the same time, projective geometry or geometry of transformations of coordinates will allow the morphology, shape and growth of living things to be summarized. The large project of a universal morphology, inaugurated by J.W. Goethe on the subject of plants[4], began to be mathematized by D'Arcy Thomson, whilst awaiting the development of differential topology. With A. Turing and his reaction–diffusion systems, the mechanics of gradient – of which R. Thom later made great use in his famous "theory of catastrophes" – began to be introduced into the theory of morphogenesis[5]. Soon, the theory of automatons by J. Von Neuman took its turn, and cybernetics by N. Wiener with his command theory and retroaction loops, which were used amongst others in the description of hormone mechanisms. In addition, Roger Buis does not leave aside the formal grammar from N. Chomsky, at the origin of L-systems by A. Lindenmayer (useful to formalize the growth of certain algae), networks from Petri, well-suited to the logical representation of certain plant morphogenesis, the direct or indirect input from quantum physicians (such as N. Bohr, M. Delbrück or E. Schrödinger) to molecular biology, the influence, also on this, of information technology, with the notion of "program", of linguistics (R. Jakobson) and of the theory of information (C. Shannon) with the notion of "code". He also collects in great detail all the inputs of structuralist thinking in mathematics which, from the theory of Eilenberg–Mac Lane categories to that of graphs and networks, have allowed a systematic and relational biology to develop, in which the notions of self-organization, emergence,

---

4 Von Goethe, J.W. (1790). *Versuch die Metamorphose der Pflanzen zu erklären*. C.W. Ettinger, Gotha. According to the author, all plants are derived from a fundamental prototype, by an action which, on the one hand, introduces diversification, and, on the other hand, collects organs around a common focal center, "in numerical proportions that are more or less fixed, but likely to be altered by circumstances".

5 Y. Bouligand was a pioneer on the subject (Bouligand, Y. (1980). *La Morphogenèse, de la biologie aux mathématiques*. Maloine-Doin, Paris). But the collective book published under the direction of P. Bourgine and A. Lesne allows measurement to be made of the path travelled from (Bourgine, P., Lesne, A. (2006). *Morphogenèse, l'origine des formes*. Belin, Paris).

complexity[6], scale invariance, order and disorder have become dominant today, leading *in fine* to the construction of biomimetic automatons (artificial life) and the development of an entire bio-informatics approach relating to simulation.

Obviously, given the spontaneous interaction of the disciplines, the existence of these empirical developments does not constitute a justification in itself. They therefore deserve to be revisited and for us to ask of them: when and under what conditions mathematics is really productive in biology? What do we expect to gain from applying it? Which mathematics should we use, where and why? Roger Buis, in the last chapter of his book, broaches these questions with courage and answers them very precisely, underlining each time the benefit that mathematics brings to biologists. If not predicting, is it about describing or explaining? Do we aim for architectures or processes? In contrast to modern philosophy, which often restricts itself, in the manner of Heraclitus, to dwell on the influence of difference, science – Roger Buis demonstrates this forcefully and epistemologists can but approve it – has the objective of finding invariants. The important thing is not that something changes. The important thing is to consider what does not change within things that change – because it is an invariant only in its connections to transformations. And we find some in biology and in physiology, as well as in physics. F. Cuvier, É. Geoffroy Saint-Hilaire and E. Haeckel already explained some. Today, we see them in metabolic cycles, macromolecules (DNA and RNA), genetic code (to the nearest few exceptions) and cell theory (J. Monod). It remains that in biomathematics, they must be linked to time and space. Since then, the use of continuous, or discrete, formalisms, of spatialized models of random or kinetic regulation processes – models such as those that Roger Buis studies competently and in detail – will contribute to their appearance. In conclusion, the singularity of living things must not be seen as an obstacle, and even though precautions and a certain modesty is required – because the model is not reality, it is, at best, only an isomorphic representation[7] – there is no doubt about the usefulness of mathematics of living things in a well-defined conceptual framework, and that it allows us to achieve the objective of all well-understood science: making sense of what we are studying.

---

6 Refer, in particular, to Monsef, Y. (1996). *Modélisation et simulation des systèmes complexes*. Technique et Documentation, Paris.

7 More often, perhaps, only 'homomorph". What is more, although the resemblance with the real phenomena has rightly been verified, it does not always carry the status of an explanation: as Loïc Forest quite correctly points out with regard to L-systems or fractal structures, the connection between reality and the model, although it may be excellent, does not mean in any way that the rules of description of the latter use real laws of plant morphogenesis. See Forest, L. (2005). Models of tissue morphogenesis from integrated cellular dynamics. Main application to radial secondary growth of conifers. Modeling and simulation. Doctoral thesis, Université Joseph-Fourier – Grenoble I, 69.

This eloquent pledge by Roger Buis in favor of biomathematics – a defense and illustration of rational models that are available – is much more than a simple memento or a catalogue. It is a true epistemological and scientific reflection, precise and nuanced, nourished in wide-reaching culture and which overcomes fractures and controversies. From Aristotle to G. Canguilhem, great names of Western thinking are found, which means that philosophy, and even the honesty of man, is not out of place. Going much further: the structure, which also reinforces the convictions of the researcher, is suitable ground for germination of new ideas. Without attempting to take on the role of a know-it-all, we even have a desire to extend it.

To mention an initial fact here, which will speak to mathematicians, today we know that non-associative algebra models Mendelian genetics, providing what is known as "genetic algebra"[8] or, more precisely, "gametic"[9]. Applied to hemoglobin, mathematics then intervenes not only as an instrument of description and explanation, but also as a forecasting instrument in research into the evolution of certain blood diseases and their oscillations. They thus allow calculation to be made of the stable states of an infected population, which corresponds to what is known in mathematics as "idempotents" of algebra, whose coefficients verify Hardy–Weinberg's law[10]. Each time the law is satisfied, a stable state for the disease will have been identified. Another example could involve generalization of the notion of an invariant. Roger Buis often highlights the need for well-defined models whose applications are valid at a certain scale, although in biology there are also multi-scale processes. But we are also aware of phenomena of scale invariants, which in particular demonstrate fractal structures. Certain forms – in particular, plants – follow these models that computer science has brought to the forefront very precisely[11]. Finally, not only what were previously known as primary qualities (above all the form), but also the secondary qualities (e.g. the color[12]), are just as easily modeled mathematically, which thus leads to the idea of a kind of "algorithmic beauty"[13] of nature. Serious ecology itself has for a long time been

---

8 Bertrand, M. (1966). *Algèbres non associatives et algèbres génétiques*. Gauthier-Villars, Paris. In addition, Roger Buis mentions them.
9 Micali, A., Revoy, P. (1986). Sur les algèbres gamétiques. *Proceedings of the Edinburgh Mathematical Society*, 29, 187–197.
10 Micali, A. (1998). Formes quadratiques de Hardy-Weinberg et algèbres de Clifford. In *Clifford Algebras and their Applications in Mathematical Physics*, Dietrich, V. *et al.* (eds.). Kluwer Academic Publishers, Dordrecht, 259–266.
11 Refer for example to Lagües, M., Lesne, A. (2003). *Invariance d'échelle, des changements d'état à la turbulence*. Belin, Paris, 328–342.
12 Refer to Berthier, S. (2000). *Les Couleurs des papillons ou l'impérative beauté, propriétés optiques des ailes de papillons*. Springer-Verlag-France, Paris.
13 Kaandorp, J.A., Kübler, J.E. (2001). *The Algorithmic Beauty of Seaweeds, Sponges, and Corals*. Springer-Verlag, Berlin/Heidelberg/New-York.

sliding from politics towards science, resolutely entering the world of modeling[14]. On balance, then mathematics no longer has the "severity" that Lautréamont lent it. Well understood, it is again gaining favor. But this is also already, more than anything else, what Roger Buis's book does.

Daniel PARROCHIA
Honorary professor
University Jean Moulin-Lyon III

---

14 Refer for example to Coquillard, P., Hill, R.C. (1997). *Modélisation et simulation d'écosystèmes, des modèles déterministes aux simulations à événements discrets*. Masson, Paris.

Introduction

> *If in biology we wish to reach understanding of the laws of life,*
> *it is then necessary not only to observe and notice vital phenomena,*
> *but in addition the intensity relations in which they are related to each other*
> *must be set up numerically.*
> *This application of mathematics to natural phenomena*
> *is the objective of all science, because the expression of the law*
> *of phenomena must always be mathematical.*
>
> C. Bernard, 1865[1]

This opinion from the physiologist C. Bernard is emblematic of the relationship that biologists must construct between experimental practice and formulation of their observations. Biology, and in particular physiology, an experimental discipline *par excellence*, is thus expressly encouraged to make use of mathematics, in its very essence an abstract science. This demanding position caught between two such different practices does not come without its problems. First, despite numerous examples that prove a positive connection between these two disciplines, its very principle sometimes causes incomprehension that is difficult to remove. Let us clearly state that some biologists do not fully appreciate the true contribution that they could obtain from mathematics, restricting themselves to using it simply to analyze their observations statistically, a given law or a given recognized model. Using it more in this manner is a calculated exercise, an illustration, than as a means of extracting new important information about properties of the processes that they are studying.

---

1 Bernard, C. (1984). *Introduction à l'étude de la médecine expérimentale*. Flammarion, Paris, 185.

Thus, establishing a connection between these two disciplines and the true benefit of combining them leaves some biologists in doubt about the practical advantage of attempting an in-depth approach to bringing them together. In their defense, it can be said that bringing them together in this way implies an "intellectual investment", which is doubtlessly more rigorous, in any case more rigorous, than the introduction of concepts and methods from chemistry and physics into biology, disciplines with the common factor of being "sciences of nature" based on the observation and measurement of tangible phenomena. Since the situation involves both combining or confronting points of view with each other and a collaboration of capabilities, we are aware of the difficulties of establishing real operational contacts, always subjected to institutional separation in both teaching and in the organization of research laboratories. Despite undeniable progress (both psychological and sociological), which must be acknowledged, here we have a recurring subject of debate that merits an overview, whilst there is an accentuation of this general evolution in which many sectors of human activity are being "mathematized". In light of this observation, it is necessary to specify that these relations are necessarily marked by what is intrinsic to each discipline in its nature and practice.

The field of mathematics is today a highly diverse body of knowledge in terms of objectives and methods. Its current state results, as we know, from a progressive extension of its various fields of study. Whilst, in Antiquity, arithmetic, Euclidean geometry and trigonometry formed its undeniable foundations, several other axioms were laid down afterwards, introducing other points of view that were particularly fruitful. As an example, we can reference the extension of the notion of number (ranging from natural whole to complex and quaternions), the move from arithmetic to algebra (notions of "group" and algebraic structures), the calculation of probabilities or even infinitesimal calculus (differential and integral, optimization), the diversity and extent of whose applications is well-known. Moreover, this evolution continues to the point that "mathematics is currently undergoing an extraordinary prosperity" in both qualitative and quantitative terms[2].

We know that there is a link between this evolution and the constant approach by mathematicians to always *refer to a given referential*, which allows them to set up the underlying layers to their work by defining an appropriate system of coordinates. According to the problems encountered, the mathematician places themselves "like a good surveyor" in a given "space", allowing them to abandon the classic Euclidian reference to work in another world, a very varied world, the specifications of which are based on the notions of vector space and topological space. A proliferation of this kind obviously raises questions for biologists from the moment that they know

---

2 Dieudonné, J. (1982). *Penser les mathématiques. Séminaire philosophie et mathématiques de l'ENS*. Le Seuil, Paris, 16.

that the processes they are studying cannot necessarily be reduced to questions of distances and metrics. But they also, depending on the case, call on questions of vicinity and limits. We are able to see in this the consequence of a sort of inhomogeneity of its workspace, which already illustrates, as we will see, the diversity of the dynamics of deterministic systems whose behavior can be marked by jumps or bifurcations – the point at which biology and mathematics comfortably coincide.

Whilst this mundane remark about the sudden appearance of new paradigms is of course true for all disciplines, we can question the status of each of them and examine what the specificity of each field of scientific knowledge would be. This is true for both epistemological assumptions and methodology, independently so from everything that arises from interdisciplinarity[3]. The question can be raised particularly for biology if we base our judgment on the permanence of the discussion surrounding the originality of this discipline with a view to specifying if and how it is different from physical sciences whilst allowing it to take root more and more[4].

In fact, these interdisciplinary relationships are often highly interlinked, in such a way that the connections that biology makes with mathematics are not always independent of those that are made with physics, where the influence of the latter lies in both the clearly more advanced formalization and a certain affinity with biology that itself has for a long time involved various physical notions at work in the different processes that it examines. Thus, there is a point of view known as "physicalistic" (or "mechanistic") that is obvious and consistent in various forms with deducing the properties of biological things from the existence of underlying physical mechanisms (or physico-chemical). This undeniably marks the positions of principle that biology attempts to establish with a view to a *formalized representation of the phenomena of living things*. In the opposite sense, a strictly

---

3 Let us recall the confusions underlined by the "Sokal affair" on a purely formal interdisciplinarity without a critical examination of conceptual analogies. Refer to Sokal, A., Bricmont, J. (1997). *Impostures intellectuelles*. Éditions Odile Jacob, Paris. On this question of analogy, we can recall the prudent words by J.C. Maxwell: "By physical analogy I mean this partial resemblance between the laws of sciences and the laws of another science, which means that one of the two sciences can be used to illustrate the other" (Maxwell, J.C. (1890). *Scientific Papers*. Cambridge University Press, Cambridge).
4 Refer, for example, to Jacrot, B. *et al.* (2006). *Physique et biologie: une interdisciplinarité complexe*. EDP Sciences, Les Ulis. For mathematics and information technology, unless an overview is given, covering a wider methodological range, their role is discussed in the context of the biology of development and of genetics in Keller, E.F. (2004). *Expliquer la vie*. Gallimard, Paris.

mathematical point of view only calls on notions or mathematical concepts without needing, at least temporarily, to attach or inject into this a solid interpretation. The question of whether this discrimination is schematic is something we are able to agree on, as attested to by the variations in vocabulary, where it has been possible to describe a single general type of approach as "biophysical" as easily as "biomathematical"[5]. By this term of "formalized representation", we understand (although the term "representation" is itself a subject of debate for certain epistemologists, speaking more about "theorization"), a *description of objects or biological processes – their specific characteristics and their properties* – that take on a "particular" language, characterized by a specific coding and rules, where writing in literary or vernacular language only constitutes an initial approach to a description, an initial report on the study.

To begin, let us lay out the objective of this book a little, using the following simple problem: how to mathematically express one of the most-researched biological processes – the kinetics of growth. Whether this is, for example, the evolution of biomass of an *in vitro* culture of cells or demographic variations of a natural population *in situ*.

A first method of expressing this, in a purely empirical way, does not lay down any *a priori* hypothesis on the phenomenon in question. It is simply associated with operation of a statistical smoothing of data by a "neutral" equation, meaning without a connection to the underlying biological mechanism (cell division, reproduction, mortality, etc.). This is the very objective of classic methods of *statistical biometrics* whose principles we recall hereafter. In the present case, we resort to a polynomial equation, for which we will need to determine the degree *n*, ensuring correct adequacy. The objective of this kind of *statistical regression* is to express the relationship between an explained variable *Y* and an explanatory variable *X* (which in this case would be time) according to the following stochastic model:

$$Y = \beta_0 + \sum_{k=1}^{n} \beta_k X^k + \varepsilon$$

---

5 Thus, the periodical *Bulletin of Biophysical Biology*, founded by N. Rashevsky in 1939, changed name in 1973, becoming the *Bulletin of Mathematical Biology*, which is still the current title and organ of the *Society of Mathematical Biology* (United States). Similarly, the famous pioneering structure by A.J. Lotka about the differential formalism applied to the dynamic of biological associations was initially called *Elements of Physical Biology* in 1925, and was then republished in 1956 under the title *Elements of Mathematical Biology*.

This model[6] only postulates the probabilistic hypotheses that define the random part ε and which correspond to the primary notion of parent population **P** or theoretical set from which we suppose that the sample (the measured Y_i) was drawn. The distribution law of **P** and its parameters (the β above) must be estimated from the sample data. The difference between this theoretical model estimated in this way and the observed data represents the part known as "residual"; differences between the observed values and the values predicted by the model. Hypotheses about **P** and the quality of estimations are the basis that is required to allow an analysis of variance and decide on the value of n, according to the risk of chosen decision.

Another approach consists of writing one or several differential equations that express the type of presumed variation f that the speed of growth undergoes over time, as a function of the biomass or of the instantaneous number, written:

$$dy(t)/dt = f(y(t), t, \mathbf{P})$$

where **P** is a set of parameters to be identified. This model can be modified by adding a delay effect that expresses the action of a previous state $g[y(t - \tau)]$, which corresponds, for example, to a period of latency or of maturation. We are, therefore, led towards formulation of specific hypotheses (that it will be necessary to verify) about these functions f and g, and about the parameters **P**, whilst *also* basing it on what we know by simple empirical observation. Each model of this type is characterized by the nature of the hypotheses set down and by the type of observations to which we resort in order to found them (e.g. the shape of the growth curves with one or several points of inflection). A model of this type is therefore mixed in nature in the sense that the hypotheses that define it are not entirely "free".

Finally, let us add in another way of proceeding that adopts an "axiomatic" position so named because it is not based on an experimental basis, but instead on *a priori* ideas that confer to it a character that is considered more theoretical or more abstract. For all this, it cannot be considered more "mathematized". Application of the *theory of automatons* to report a growth or morphogenesis correctly illustrates this position of principle. In this case, we lay down as point of departure the vocabulary of the different possible states and the rules of transition (generation of a new cell, change of state). We will see that the difference with the differential

---

[6] We can refine this principle of simple regression using the method of orthogonal polynomials, well suited to determination of the minimal degree n of this regression. See Buis, R. (2016). *Biomathématiques de la croissance*. EDP Sciences, Les Ulis, detailed method with example in chapter G of the online companion volume to that book.

formalism is not reduced just to the choice of continuous *versus* discrete, but affects the adopted hypotheses.

This type of discrimination between different types of models (equivalent *grosso modo* to opposing something purely formal with something that is partly empirical) of course allows us to add a little order to a varied set of approaches[7]. However, these distinctions are rather conventional, meaning by this that the essential consists instead of clearly resting on the three following points:

i) The underlying motivation (what are we looking for, where do we want to get to?);

ii) The contribution from hypotheses (on what basis?) and observations (of what kind?);

iii) The type of mathematical formalization used: its originality and its constraints or limits.

Moreover, this is not without its link to the traditional debate about the nature of mathematics themselves. These, as we know, are not independent of considerations of an experimental nature that have contributed to their own development in parallel with their fundamental axiomatics. The subject is thus rich with a variety of positions, leading to underlining the characteristics, and taking into account the historical and epistemological aspects of these connections.

These aspects will be addressed later concerning the essential "points of view" of the biomathematical panorama, which is still in the process of being established and justified. But we can agree that this isn't the place to talk about developments and must refer back to more specialized studies when a given fundamental situation relating to philosophy of biology takes shape. At least it seems useful for us that this place allows the diversity of situations to appear by trying to highlight the pertinence and the particularities of representation that each of the considered approaches automatically offers. This difficult task is not in vain, as the course of biomathematical literature allows distinction to be made between what still comes from simple speculations (like a kind of expedition to develop) and what, given a true effectiveness, presents itself from now on as a true instrument (both technical and conceptual) that the biologist can use for the benefit of their own work. In any case, we believe that these conditions must define the complexity of these biological–mathematical relationships, and must thus seize the originality of different approaches, which it would be better to specify without exaggerating their differences, as then it will not lead to irreducibility. In a rather abrupt illustrative manner, we could say that in this

---

7 Varenne, F. (2010). *Formaliser le vivant : lois, théories, modèles ?* Hermann, Paris.

exploration, there are numerous points of view, yet not all their sources can be easily used. Thus, we can contribute in this discussion to the reasoned expectation of biologists confronted by the continuous extension of the place of mathematics, itself highly varied, in practice of his discipline.

1

On the Status of Biology: On the Definition of Life

Stating that biology is the "study of living things", or processes that take place in the latter, comfortably sidelines what is meant by the "life" part of this term, which is supposedly their *sui generis* characteristic in comparison to "non-living things". Without doubt we can make the observation that the definition of life can only be an ideal point of view, otherwise seen by some as useless speculation. However, insistence on the question demonstrates the originality of biology in comparison to other disciplines that do not appear to have this type of requirement. Physics, for example, does not need to "define" matter or energy in order to study the phenomena in which they intervene, giving them instead the role of "variables" that can be manipulated, whether theoretically or experimentally. Let us recall the basics of the fundamental concept of force, introduced in Newton's second law, using which we can deduce a satisfactory definition of it by multiplying the mass and the acceleration of a movement[1].

Let us return to C. Bernard, who provides an explanation of the notion of life. Stating that "physiology is an experimental science [that] has no place giving a priori definitions", he considered that "it is illusory and irrational, contrary to the very spirit of science which is to seek an absolute definition"[2]. In fact, we can state that biology ignores the notion of life, whereas this is paradoxically the purpose of its field of study. In summary, he goes as far as saying that the field refers to objects that common sense describes as "living". But this does not prevent the definition of

---

1 Nevertheless, let us note that Poincaré considered the criss-crossing, the vicious circle, to be inextricable and to consist of defining the mass by the force or *vice-versa*. Poincaré, H. (1968). *La Science et l'Hypothèse*. Flammarion, Paris, 118.
2 Bernard, C. (1885). *Leçons sur les phénomènes de la vie communs aux animaux et aux végétaux*, vol. 1. Librairie J.-B. Baillère et fils, Paris.

"living things" (as opposed to "life" itself) – moreover, we prefer to talk about singularity rather than look for a true definition of it – remaining a theme of reflection, the subject of university PhDs, stimulated by the latest developments in research on the initial conditions of appearance of life, as well as sometimes by metaphysical considerations that we do not need to include here.

Unfortunately, this pragmatism can only sideline, without resolving, the remaining reason for the existence of the debate. There is, in fact, a debate to be had if we note the diversity of the points of view, ranging, for example, from the statement by the biochemist Szent-Györgyi that "life as such does not exist", to attempts at a definition, if not about the word "life" designating a specified entity, then at least about the nature of what makes an organism "living". Although Aristotle proposed a fundamental definition of it, based on a property of autonomy and self-reproduction (see Chapter 3), modern biology considers that life is an abstract entity that it is unable to characterize. François Jacob expresses this in the following way:

> "Life is a process, an organization of matter. […] We can therefore study the process or the organization, but not the abstract idea of life. We can attempt to describe and we can attempt to define what a living and a non-living organism is. But there is no 'living matter'"[3].

Let us agree that the designation "living matter", refused in the above, is simply a convenience of language, like the famous "vital force" advocated by vitalists. And, since it is necessary to take a side, let us admit the following position concerning our subject, which will be illustrated several times over in the following.

DEFINITION.– Any living object is a precise assembly of interactive "elements" (meaning matter, energy, information), and a system of this kind always faces a risk of instability. It is therefore necessary to search for the structural characteristics and dynamic properties that ensure correct maintenance, development and reproduction of it.

Faced with this issue, mathematics, which is itself a science of structures and transformations, steps up to participate in this study and, better still, to establish a coherence of dynamic representation, in part invariant and in part fluctuating (adaptation).

With this in mind, we can refer to the principle stated by H. Atlan (and independent of its position with regard to the deterministic/random debate): "The only specificity of living things relates to the complexity of their organization and

---

3 Jacob, F. (2000). *Qu'est-ce que la vie ?* Éditions Odile Jacob, Paris.

the activities that accompany them"[4]. Obviously, we need to specify this notion of complexity and distinguish it from what it is in non-living systems (Chapter 3). In the meantime, let us say that the notion of life, understood in this way, consists of admitting the following property.

PROPERTY.– The most characteristic processes of any living thing, which can be described as *sui generis*, are *autopoiesis* (continuity of an autonomous production of oneself) and *compliant reproduction* from one generation to the next.

Without spending too much time on this question, we at least need to agree on what and how biology presents a demonstrated specificity with respect to other disciplines, leading us to say, in the words of the biologist E. Mayr, that it is "a science unlike any other"[5]. We call on some past considerations in order to do this.

Let us first recall that the word "biology" is relatively belated, dating back only to 1802, proposed jointly by J.-B. de Lamarck in France and G.R. Treviranus in Germany. For G.R. Treviranus, "biology or philosophy of living nature" ("Biologie oder Philosophie der lebenden Natur") must aim to study "various phenomena and forms of life, the conditions and laws that dictate its existence and the causes that determine its activity"[6]. For his part, J.-B de Lamarck entitled one of his lessons "Biology or considerations on nature, faculties, developments and the origin of living bodies" (1812). Moreover, he specified:

> "The name of living bodies has been given to these singular and truly admirable bodies... They effectively offer, in themselves and in the various phenomena that they present, the materials of a particular science that has not yet been founded [...] and that I shall name biology"[7].

With this new terminology, the objective appears first as the desire to group together all the studies, scattered to a greater or lesser extent, that relate to living things, going beyond the observation and classification of living things whose morphology was the basis for recognition, naming and classification. This first requirement did not prevent a reasoned practice of the use of living things developing in parallel within this natural history, duly distinguished by their nature and their state of development (pharmacopeia, feeding, clothing, housing). This implied a technical experimentation that should not at all be considered negligible

---

4 Refer to Stewart, J. (2008). La Vie existe-t-elle ? Déterminismes et complexités: du physique à l'éthique. In *Colloque Cerisy*. La Découverte, Paris, 145–158.
5 Mayr, E. (2004). *What Makes Biology Unique?*. Cambridge University Press, Cambridge.
6 Quoted in *Encyclopedia Universalis*.
7 *Ibid.*

(harvesting and hunting, then agriculture and domestication). But the most important point in this "profession of faith", namely biology, is now to consider life as a singular phenomenon whose two aspects are specified by it as: its generality and its diversity. A two-fold task is therefore assigned to this new discipline: (i) looking for properties or characteristic laws that have a certain universal value and (ii) studying the diversity of how these are manifested in reality, including the truly historical aspects[8].

The eruption of the word "biology" is often (wrongly) considered as the declaration of the status of this discipline as a true experimental science. At least, this was written by F. Magendie (1783–1855), in reference to Galileo, writing in his era in his *Principe élémentaire de physiologie* [An Elementary Treatise on Human Physiology] (1833): "in order to know nature […] it was necessary to observe and above all interrogate it via experiments". This position was taken up and developed by C. Bernard, who studied the lessons given by Magendie, and whose successor he became at the Collège de France.

In fact, it must not be forgotten that experimentation on living things has been around for a long time. Let us recall a few notable examples. In Antiquity, we have Hippocrates and his experimental studies on the development of a hen's egg. Then the 17th Century provides us with remarkable research by W. Harper (1578–1657) on blood circulation. An important point is that the originality of these studies was the methodical implementation of quantitative measures of blood flow under different conditions (application of a tourniquet, then its release). Let us note also in this era the tests by L. Spallanzani (1729–1799) that led to the disproving of the theory of spontaneous generation and the highlighting of the role of gastric juices in digestion.

Natural history therefore realised very early on that the anatomical description of structures must precede research into their functioning, without remaining limited to questions of compared morphology and classification, which led to progressive labeling of the reality of various fields of study of phenomena involving living things.

1.1. Causality in biology

On the sidelines of this fundamental topic, it is necessary to mention an old question that has been around in the field of biology for a long time, in

---

8 On this point, interested readers should refer to Gayon, J. (2004). De la biologie comme science historique. *Sens public* [Online]. Available at: https://www.sens-public.org/IMG/pdf/SensPublic_Jean_Gayon_Biologie.pdf.

particular in embryology, in reference to whether the development of an organism consists of deployment of pre-existing structures in the egg, or, on the contrary, whether ontogenesis takes place according to a spatially and temporally organized sequence of generating processes. This was the famous dilemma of "preformation versus epigenesis". We all have in mind certain famous figures from the past in anthropology, illustrated representations of a miniature being (*homunculus*) that is thought to be housed either in the head of the spermatozoid, or in the ovum. This debate was already alive in the era of Aristotle, who was a supporter of epigenesis or the progressive formation of a series of structures. For a long period of time, this position remained up to date, fed by various staunch positions in which naturalists and philosophers played a part, as well as physicians such as Maupertuis who, supporting his ideas with interbreeding and cross-breeding, was firmly opposed to any idea of preformation. The issue is significant because attributing a causal nature to the notion of preformation is the same as being opposed, automatically, to both reproduction and evolution.

Although the debate eased off with the advent of cellular theory at the end of the 17th Century, which designated the cell as the elementary level where we can site both reproductive activity (mitosis) and various morphogenetic and physiological processes, the theory of preformation is still present in biology. It is of course a renewed form, but one that should be mentioned in relation to the general principle of causality. A first aspect of this persistence can currently be seen in the significance that it is often necessary to apply to the initial conditions of a morphogenetic process. In fact, this consists of taking into account the initial state of a cell at the time of its formation, as a material cause (in Aristotle's sense) of a process. Although this idea of predetermination should be distinguished from the deployment of a pre-existing structure, the initial state is indeed equivalent to a pre-requisite driving factor that participates (in part) in the advancement of a process. Thus, any change in the initial state can modify, both qualitatively and quantitatively, the dynamic of the aforementioned process. Another aspect, amongst the most remarkable, was provided to us by genetics as soon as this broke free from its initial principle of direct, linear and unidirectional causality, from gene to function. The discovery of development genes (with the existence of homeobox genes) in fact underlines this fundamental property of the genome that is to possess a complex topological and temporal structure, non-linear in its organization and its operation. It follows that the ontogenesis of an organism needs to be related to this "preformed structure" whose temporal evolution takes on the meaning of a multidimensional "deployment", not of the genome itself as a physical object, but more exactly of the information that is written in it. Which can be translated by saying that the "homunculus is indeed represented on chromosomes"[9].

To organize the subject, we first need to sum up the position of the principle of causality, the basis for all scientific knowledge. Although it is used very frequently, the debate continues around what is meant by "causality". Let us sum up by saying that, in a more general manner, the function of this term can be given as follows: "reconstituting links, adding events that at first glance are disjointed into a timeframe"[10]. It is a search for an ordered sequence of elements that are linked not necessarily by immediate proximity (spatial or chronological), but by the non-circumstantial fact that they find themselves together and in a given orientation on a single arrow of time, one which allows a given process to be structured. This is illustrated by representation on a directed graph, connecting two elements of the same set, covering the immediate direct effects just as much as the delayed effects (such as an incubation period or a dormancy period) or retroactions (feedback).

This point of view, distinguishing causality and correlation, conveys an ontological character to the idea of causality: the succession of phenomena over time necessarily determines what happens during the evolution of a being or of a system[11]. Remaining on this theme, the question arises of whether there is a link between the proven existence of causality and the explanatory nature of our concepts and models. Some advocate that there is, believing that a mathematical model works (in the sense of simulation) because it explains even a little of the information that participates in causality. Now this needs to be explained further, because if a model is capable of bringing characteristic properties, the latter can come from a simple statistical correlation or from a causality. Moreover, we will see that in terms of modeling, we can be brought to make a mathematical distinction between a deterministic part and a random part, respectively corresponding to what we study in terms of inventoried causes and what we do not know about in terms of unknown or inaccessible causes.

We are aware of how much the natural sciences, including biology, were influenced for a long time by the thoughts of R. Descartes, who stated that it is necessary to move "from what is most simple to what is most complex in the order of deduction" on the basis of "these long chains of very simple and easy reasons that geometricians make use of in their proofs"[12]. The principle of these "chains of

---

9 Prochiantz, A. (1994). *Forme et Croissance*. Le Seuil, Paris, 19. French translation of Thompson, D'A. (1992). *On Growth and Form*. Cambridge University Press, Cambridge.
10 Andler, D., Fagot-Largeault, A., Saint-Sernin, B. (2002). *Philosophie des sciences*, vol. 2. Gallimard, Paris, 825.
11 Refer to Kant's distinction between necessary causality and free causality, meaning between determinism and freedom.
12 Descartes, R. (1966). *Discours de la méthode*. Garnier-Flammarion, Paris, 47–48.

reasoning", unidirectional linear sequences, is relayed today by the preeminence given in biology to the circular causality linking in both directions two entities that are at a lesser or greater distance from each other, which complete or modify the direct linear causality[13]. In ecology, this notion of circular causality was proposed for the first time (by Hutchinson in 1948) without the specific connections with the principles of cybernetics that were developed in the same era by N. Wiener having yet been established. This *circular causality* takes on different forms up to the remarkable existence of loops (e.g. metabolics) or open circuits such as those in the *hypercycle* by M. Eigen. It is clear that currently the notion of causality must be designed as much as possible in the context of networks, meaning directed graphs connecting various elements that participate in the same type of process (see Chapter 3).

In a general manner, it is sensible to connect these considerations to the usual and still relevant distinctions accepted since the time of Aristotle between various types of causality[14], in particular the efficient cause and material cause. This distinction has been taken up by the bio-theoretician R. Rosen (see section 3.12.2) in applying it to the classic representation of a biological process with the aid of a dynamic system, meaning a set of differential equations differentiated by time, $dy(t)/dt$, expressing the speed of the aforementioned process. A speed equation corresponds to an efficient (or formal) cause, since its formalism directly determines the advancement of the process. Now, this also depends in general on the initial conditions $y(t=0)$, where these can have a greater or lesser influence on the dynamic of the process. Let us specify, without waiting for later developments, this question of dependence with respect to the initial conditions by highlighting that it cannot be strictly quantitative by simply modifying the order of magnitude of the final state. It can also be qualitative. Indeed in certain cases, the initial state can condition the very type of dynamic as the evolution towards a particular "basin of attraction" (case of multistationary systems). This is, for example, the case for certain phase transitions, like in plants the maintenance of a meristem in a vegetative state versus its evolution towards a reproductive state (no longer generating leaves but flowers instead). The initial state then does indeed have the meaning of a cause, known as a matter cause. This means distinguishing, by analogy with mechanics, the movement of a body determined by a formal cause (its speed equation) and by its presence at the initial time (*material cause*). For a cell

---

13 Concerning this question, refer, for example, to Mossio, M., Bich, L. (2014). Biological circularity: concept and models. In *Modéliser et simuler*, vol. 2, Varenne, F., Silberstein, M. (eds). Éditions matériologiques, Paris, 137–169. The general theme of causality was the subject of a thematic school of thought within the CNRS (French National Research Institution), entitled "*Corrélation, causalité et régulation en biologie*", Île de Berder, 2013.

14 The four types of causality according to Aristotle are the causes known as "material", "formal", "efficient" and "final".

growth process, we have something equivalent when we distinguish between generation (local mitotic activity that generates a new cell of a given size and in a given state) and increases in size (subsequent activity of elongation of the wall).

Apical growth of a mycelial or algal filament if a good illustration of this, since the sub-apical cell can, depending on its initial state (independent of other causes), either grow without differentiation and participate in extension of the filament in the same direction, or be at the origin of a morphogenesis by budding and generating a lateral offshoot (ramification).

In the elucidation of the relationship that is required to exist between two phenomena and which can be considered to be of a causal nature (and not simply a correlation), two concepts need to be mentioned due to the epistemological importance that they had or still have in biology: vitalism and teleology. We have seen them, albeit in an exaggerated manner, as past obstacles to biology reaching a status of science that can be compared to physics[15]. It rather appears to us that they simply reflect the difficulty of agreeing on the very notion of life.

1.1.1. *Vitalism*

By vitalism, we mean a position of principle that aims to make a fundamental distinction between living things and physical material objects. For Descartes, for example, a living organism is nothing other than a machine, where no difference is seen between living matter and matter known as "inanimate". On the other hand, vitalism considers that living matter has specific properties that cannot be reduced to the set of physico-chemical elementary mechanisms, from which we deduce that they must result from the existence of a "life force" (*vis vitalis*) that is specific to it.

In principle, this concept dates back to Aristotle, with his definition of living bodies (*On the Soul*, volume II, 1). By "life", he clearly meant "the fact that they feed, grow and decay *by themselves"* (our emphasise). He thus expressed a principle of autonomy as a fundamental characteristic of life, meaning something intrinsic that inanimate bodies do not have. In a way that is more practical than philosophical, the idea of vitalism was laid out by Galen (129–~200). According to him, the study of structures and functions cannot be separated from the notion of utility as the very objective of life[16]. It somehow combines determinism and finality. Following this, vitalism was illustrated in different ways, in particular by the embryologist H. Driesch (1867–1941) in terms of the theory of epigenesis, in contrast

---

15 Mayr, E. (2004). *What makes Biology unique?*. Cambridge University Press, Cambridge.
16 Galien, C. (n.d.). *De l'utilité des parties du corps humain*. See Pichot, A. (1993). *Histoire de la notion de vie*. Gallimard, Paris, 130 *sq*.

to the idea of preformation. The philosopher H. Bergson (1859–1941) actively participated in this movement in his relationships with evolution. His expression "vital impetus" (*élan vital*) means the existence of a metaphysical principle that is established at the very origin of life, the principle of "given once and for all", which then maintains itself as a connecting link between generations (*L'évolution créatrice* (Creative Evolution), 1907). In this concept, H. Bergson sees, in his own terms, "a limited force, that always seeks to surpass itself, and always remains inadequate at the work that it attempts to produce"[17].

If the existence of a vital impetus cannot be demonstrated as a defined and manipulable entity, the debate about vitalism allows us to shed light laboriously on the requirement that living mechanisms should not be limited to just the elementary laws that govern physico-chemical phenomena. Thus, with C. Bernard, whose dialectics are known, we can reject the idea of a vital principle whilst conceding alongside him that "It is obvious that living beings, by their evolutive and regenerative nature, radically differ from non-living entities, and in this respect, it is necessary to agree with vitalists"[18]. The doctor and contemporary philosopher G. Canguilhem follows the same path.

Let us explore this question of vitalism a little by noting an important point that came to light when we became aware that there was definitely a "living thing-environment" set that could constitute a particular significant entity, a "totality", the terms of which cannot be dissociated in the study of certain phenomena. The biologist J.J. von Uexküll (1934) created the term *Umwelt* to designate the behavioral environment specific to an organism[19], which needs to be distinguished from the strict topographical environment. Each organism has an environment that is well specified *lato sensu* with which it is in a close relationship. Although this concept usually refers to the sensory environment of animal species, from an ethological point of view, it is necessary to extend it to the study of other organisms where we know that various tactisms and tropisms occur, as necessary constitutive deciding factors of their environment.

Let us note that this is part of a wider set of considerations about the lack of reproducibility of the dynamic of a given process when moving from an *in vitro* experimentation to a more integrated *in situ* study. An example is given to us with

---

17 A comment is made about this by P.-A Miquel (Miquel, P.-A. (2007). *Qu'est-ce que la vie ?*. Vrin, Paris), who insists on two points: (i) life is power: "it is made by it"; (ii) and yet it is "inadequate for its work" since it is subject to a limit that is essential (internal) and non-accidental in nature.
18 Bernard, C. (1867). *Rapport sur les progrès de la physiologie en France*. Imprimerie impériale, Paris.
19 Refer to Canguilhem, G. (1980). *La Connaissance de la vie*. Vrin, Paris, 143 *sq*., who underlines the similarities with *Gestaltheorie*.

the case of plant cell growth (see Figure 3.11). Indeed, control of parietal extension leads to intervention *in situ* of the electrostatic potential of the wall, because this modulates the affected enzyme kinetics in terms of their microscopic properties exhibited *in vitro* (and qualified as intrinsic). It is clear that these are not sufficient for a correct representation of the dynamics of the growth process[20]. More generally, therefore let us say:

PROPERTY.– The determining set of interactions of a biological system with its environment constitutes an original characteristic of living things. This systemic view of nature is specific to biology, without an equivalent in physical sciences. In other terms, this is the question of the existence of a fundamental environment: dependence that is continuously in place in all ontogenesis.

With this mindset, the biologist R.C. Lewontin talks about a triple helix, a metaphor that is intended to outline, as if on an equal footing, a triple determinism: genome, organism and environment[21].

1.1.2. *Teleology*

Teleology can be defined as a study of the finality that can be attributed to any phenomenon. It lays down the principle of the existence of "final causes", using Aristotle's distinction between the existence referred to in the above of several types of causality, in particular an efficient or immediate cause (a phenomenon that produces another) and final cause (objective of the action). Obviously, we can reason in a consciously anthropomorphological manner, saying that having an objective constitutes in itself the determinism of all actions. More simply, looking here from the point of view of the "economy of life", we will say that it comes down to asking the question (for coherence reasons, inescapable): "What is the use of this?". Otherwise expressed, it is a case of elucidating what the "relationships between means and ends"[22] are, and not becoming fixed on just the "common sense" that says it is possible to predict the present from the past (that is known), but not the future (that remains unknown), meaning that the cause precedes the effect.

---

20 Ricard, J. (1990). Le fonctionnement des enzymes en milieu cellulaire. *La Vie des Sciences*, 7(3), 197–218.
21 Lewontin, R.C. (2003). *La Triple Hélice: les gènes, l'organisme, l'environnement*. Le Seuil, Paris. Lewontin's book refers to the genetics of populations and the theory of evolution, and his research is marked by its use of mathematical formalism whilst developing its own philosophical thinking.
22 According to the definition in Lalande, A. (1968). *Vocabulaire technique et critique de la Philosophie*. PUF, Paris.

Teleology often has bad press amongst biologists, despite the essentially metaphorical nature of this term. In terms of evolution, talking about teleology comes down to attributing an objective to a particular set of evolutive processes. Thus, concerning the formation of organic structures that provide vision, from initial rudimentary photosensitive elements to the complex eyes of superior animals, it can be said that "the eye is made for you".

On this point, C. Darwin himself recognized that it was absurd to believe that the eye, with all its adjustment devices, was able to form itself solely through natural selection via a series of evolutive processes. From the teleological point of view, this evolutive convergence must be interpreted in fact as an optimization of the function of vision. Ontologically, the principle of teleology means that the development of a living thing, from an embryo to an adult state, has no "objective" other than compliant production of physiology and morphology that are characteristic of the species in question, adjusted to a lesser or greater extent by environmental constraints (morphogenetic plasticity). Anything that is simply a tautology obviously disappears and presents a problem when we look at how this remarkable production functions, with its certain stability or property of invariance.

Opposition to the very idea of teleology consists of refusing a final cause that would have an explanatory value for life processes[23]. Certain biologists wanted to rid themselves of this cumbersome term with a finalistic connotation by talking instead about "teleonomy". In fact, this changes nothing in the essence of the debate since the two terms have the same etymology (*teleos* = purpose, objective). This designation of teleonomy was proposed in 1958 by the biologist C.S. Pittenburgh who worked on the circadian rhythms of drosophila. It was taken up by the neo-Darwinist E. Mayr, who defined it as an unintentional outcome ("non-purposed end-seeking process"), an expression that is very difficult to define. Excluding any quibbles from a language point of view, F. Jacob and J. Monod use this term to mark out the principle of final cause by laying down a new founding principle: living things are "objects equipped with a project". In place of the principle of an explicit final cause, we therefore substitute the idea of a program or a series of determining instructions which link structure and function.

F. Jacob insists particularly on this new concept of program that he sees as the possibility of getting rid of old notions of finality and mechanisms. Moreover, he notes (*The Logic of Life*, p. 18) that the idea was already implicitly present in C. Bernard, whom we partially quote below:

---

23 Of course, here we are not referring to the idea of "intelligent intent" in the sense that a point of view of a metaphysical order does not arise from this current research.

"There is something like a pre-established intent of each being and each organ, so that if, considered in isolation, each phenomenon of the economy is a tributary of the general forces of nature, taken in its relationships with the others, it reveals a special link, it seems directed by some invisible guide in the road that it follows and brought to the place that is occupies"[24].

Of course, in his era, these terms like "pre-established intent" and "guide" thus laid down had to appear in a teleological manner. But we know that the dialectics of C. Bernard led him to admit pragmatically, and also with a sense of nuances that is sometimes fortunately missing from the discussions, that contradictory positions exist (see Chapter 3).

A new debate begins concerning the notion of a "program" which, with its connotations of data processing, has the meaning of an algorithm that codes and assembles the basic information obtained by molecular biology. On this point, advances went up to the point of saying that the keyword of biology was no longer "organization" or "organism", but "information". "Besides the notions of energy and mass in physics, reaction and stereospecificity in chemistry, biology used from then on words of information technology and programming[25]". In any case, to remain as close as possible to the notion of causality that we are discussing, let us say that if the gene *lato sensu* can be a direct cause producing a given effect, it does so in the form of a coded instruction or program procedure. We will come back to this duality.

This epistemological evolution leads to the consideration that the originality of living things, in comparison with inanimate things, lies in the existence of a kind of double causality: every biological process is subject to specific quantitative laws (just like any physical phenomenon) at the same time as being determined and controlled by genetic programs. However, and without waiting for our later remarks on the notion known as the "complexity" of biological processes, the question is not closed, because the very term causality should be explained due to another important characteristic of biological phenomena; their variability, whose properties we must examine.

---

24 Bernard, C. (1879). *Leçons sur les phénomènes de la vie communs aux animaux et aux végétaux*, t. 2. Librairie J.-B. Baillère et fils, Paris.
25 Maurel, M.-C., Miquel, P.-A. (2001). *Programme génétique: concept biologique ou métaphore ?*. Éditions Kimé, Paris, 41.

1.2. Variability in biology

We first note the underlying difference that exists between a chemical species, easily manipulated and reproducible in an identical manner, and a biological species, an entity that even in the simplest cases of genetic lineage has a random part, including the fact that the genome is not strictly invariant from one individual to another within the same species. A precise example is that of the immunological characteristics of allotypes and idiotypes concerning the individual specificity of immunoglobins, or even the existence of blood groups in mammals. On the other hand, we know of the importance of randomness in certain biological phenomena, such as sexual reproduction (e.g. the random nature of pollination in allogamous plants), occurrence of mutations, replication errors in the genome or even chromosomal crossing-over.

Due to the importance of this notion of variability, we must specify the meaning of this term, which is used in varied situations. Depending on the case, we can consider it from an ecological and phylogenic point of view or, on the contrary, envisage it in relation to the ontogenesis of a given species. On this last point, we know that certain macroscopic characteristics of an organism present themselves as a matter of fact as random variables. For example, for a given plant species and cultivar, the number of metamers of a stem of an annual plant obtained at the end of development can vary significantly within the same population. On the contrary, although the cases are quite rare, this number is obviously constant in certain highly selective species. We observe, for example, in wheat, that the main stem includes seven internodes, with little variation of this number, but this is not repeated on the organism itself in its entirety including branching. Another example in this order of ideas is that in *Arabidopsis thaliana*, the model plant of molecular biology, on certain mutants and for a given light, a constant number of leaves are generated before floral induction. Another aspect, at the scale of a population, is that the statistical variability of the size of a given organ does itself vary during growth. For example, the length of the hypocotyl (= 1st internode of seedling) presents a coefficient of highly fluctuating variation (standard deviation/average). Therefore, in the papilionaceous plant *Lupinus albus*, this statistical index fluctuates from 5 to 60% depending on the stage of growth.

We can currently specify various elementary causes of variability. Let us summarize the two microscopic levels where they take place. An initial source consists of the spontaneous and random occurrence of genetic mutations (local alterations in the sequence of nucleotides that have repercussions in the metabolic functions that follow). These are alterations of a qualitative nature, considered to be independent one after the other without an adaptative value. On the same subject, another cause of variability lies directly in a stochastic expression of genes.

The existence of random phenomena here has the complete properties of a noise, in the sense that it is a case of quantitative (and not qualitative) temporal variations of proteosynthesis.

A second type of variation is added to this, known as "epigenetics" since it occurs outside gene determinism. We know that this genetic/epigenetic distinction is delicate. Effectively, since the discovery of the lactose operon (see Chapter 3), meaning a set of genes that intervene for the same metabolic function, the idea of the existence of interactions between genes and genetic regulation networks has emerged. The term epigenetics needs to be reserved for something else, in particular the modification of chromatin (DNA-proteins association) by methylation of histone proteins. Such changes, which are indeed epigenetic in nature since they operate outside the genes themselves, are not fixed, whereas mutations are stable modifications subject to the phenomenon of heredity. To widen this table of causes of variability, let us mention, at another level, the intervention of randomness in certain enzymatic reactions, like in the case of reactions with two substrates without privileged fixing of the enzyme onto one of them.

From a more general point of view, the significance of biological variability was well elucidated in its time by the zoologist and biometrician G. Teissier[26], based on his research using cultures of micro-organisms *in vitro* in controlled experimental conditions. Whilst this type of rigorous experimental protocol allows fluctuations from one test repetition to another to be reduced, it allows significant variability to persist as something that is, let us say, "intrinsic" in nature. It is not simply a question of precision of physical measures, but of the existence of a large number of factors, of which the experimenter can only take into account a very limited number. Therefore, let us imagine the well-known difficulty of an exact reproducibility of biological tests, because mastery of operating conditions is never total. A very small change in an experimental protocol, which can even take place without the experimenter's knowledge, can significantly modify the result, a result of the sensitivity of a living thing to its environment, to which it adapts, and due to which it modifies its response to a greater or lesser extent.

We encounter this type of question when we resort to mathematical models of test plans (experimental designs) (Chapter 3). These are always stochastic models whose random parts represent all factors that are unknown and those that cannot be controlled in experiments, without necessarily relating them to the uncertainty of measurements.

---

26 Teissier, G. (1936). La description mathématique de faits biologiques. *Revue de métaphysique et de morale.* 43(1), 56–58.

Currently, the question of biological variability is asked in a more precise manner by considering both time and the various levels of organization. This multi-scale temporal approach is particularly interesting in development biology. For example, during early development of sea urchins, which is one of the organisms that has been well-researched with this in mind, inter-individual variability of the dynamic of cell proliferation varies according to the scale considered, high over the entire population and lower amongst groups of cells of the same type and of the same generation[27]. This being so, we will return to and develop the principle of stratification that is well known in population statistics.

Whatever the source of variability from amongst the summary that we have just done of it, the part played by random nature in living things is incomparable to its situation in the physical world, where research on it constitutes the well-outlined field of physical statistics. We recall that this developed from the kinetic theory of gases in light of the macroscopic description of a set of many microscopic constituents with random behavior (molecules, atoms, ions and particles) (refer to the example of Brownian motion).

It follows that biology can scarcely be said to have a set of duly inventoried laws that are analogous, in their rigor and in the outline of their field of action, to that of physics. Nevertheless, we will see some remarkable cases of biological laws (Chapter 4).

But other considerations need to be taken into account to understand how this intrinsic variability affects the perception that the biologist has of living things, thus entailing close interest in the relations between biology and mathematics. Essentially, as mentioned previously, it is the case that objects and biological processes are generally time-dependent, on the one hand, and environment-dependent, on the other. Let us detail a little further what this entails.

1.2.1. *Time-dependence of biological processes*

At any organizational level that we choose to place ourselves, all processes that take place in living things present this characteristic. Let us give some convincing examples of this outside the realm of evolution where, by definition, biology is a historical science.

Dynamics of living things are never strictly punctual. We see this, at a molecular level, with homeostatic variations of the internal environment, generally subject (except for all pathologies) to fluctuations over the course of a nychthemeron

---

[27] Refer to Villoutreix, P. (2014). Vers un modèle multi-échelle de la variabilité biologique. In *Modéliser et simuler*, vol. 2. Varenne, F., Silberstein, M. (eds.). Éditions matériologiques, Paris, 660 *sq*.

(e.g. variations in glycemia), in such a way that it is more exactly a case of *homeodynamics*. This made C. Bernard mistrustful of average statistical values, because they were hiding variational properties that should not be neglected. Chronobiology came to life on the basis of this banal statement. More generally, this corresponds to the following characteristic dynamic of a living thing:

PROPERTY.– Independent of the fluctuations in the environment, there are relatively few punctual attractors that lead to a unique stable stationary state, but more often cyclic attractors (e.g. a limit cycle), where these can themselves fluctuate ("strange attractors"). Added to this there is the case of multistationary systems, for which there is a possibility of bifurcation or a qualitative change of dynamic.

Although dynamic properties of this type are not specific to biology (e.g. the chaotic dynamic of physical complex systems), we can state that physics concerns bodies or objects that are invariant with respect to time, with the exception of the well-described phenomena of hysteresis (memory), relaxation or radioactivity. This was highlighted by the physician Delbrück when he took an interest in genetics, declaring himself surprised to observe the absence of "absolute phenomena" in biology where, he noted, "everything depends on time and place".

On the other hand, at different organizational levels, living things are constantly subject to molecular (metabolic *turn-over*), cellular and organic (regeneration) processes of renewal or repair. An analogous behavior is demonstrated in the dynamics of populations and ecosystems, in particular with genetic mixing in meiosis (mitosis of sexual cells going from a ploidy of $2n$ to n chromosomes) and therefore the dynamic of genotypes within a given species.

Plants make up a very remarkable case which, even in annual species, are generally subject to continuous embryogenesis throughout their existence by maintaining meristematic activity (generation of new metamers by different buds, formation of conducting tissues by the cambium of roots and stems). These processes of neoformation are associated with diverse renewals of tissue and organics (sclerosis of old wood or duramen, replacement following natural abscissions, leaves and branches, morphological reiterations in arborescent species). Thus, we can state that the life of a plant is confused with its growth[28], and that the higher plant must be seen as a metapopulation. Development of certain organisms in colonies (such as corals) is another example of this, as well as, more generally, the process of aging and death, where the latter is consubstantial to life[29].

---

28 According to the expression by Hallé, F. (1999). *Éloge de la plante*. Le Seuil, Paris.
29 The famous definition by M.F.X. Bichat (1799): "Life is the set of functions that resist death" was reviewed by H. Atlan (1979), for whom life was characterized by "functions capable of using death".

Underlining these characteristics of time-dependent biology will lead us to mention the notion of time in biology, with the search for a time described as biological (Chapter 5), different from sidereal time.

1.2.2. *Environment-dependence of biological processes*

In addition to this dependence on time and on the past, objects and biological processes are, we have said, under the influence of their environment in the wider sense of the term. This is demonstrated in the property of adaptation that is so characteristic of living things. By "adaptation", we mean the operational adjustment, both morphogenetic and physiological, to fluctuations of the environment. We will talk about this in the context of the autonomy of living systems, which is found just as much in the being ("self") as in the surrounding environment.

The idea of an influence of the environment on living things is often attributed to Lamarck who believed that he saw in this the basis for inheritance of acquired characteristics. In fact, the role of the environment dates back to I. Newton, with the former physical concept of ether, whereas for R. Descartes, for example, the notion of the environment does not exist, and all physical action comes from direct contact. For I. Newton, ether places various objects in continuity. He attributed the understanding of the physiological phenomenon of vision to it. He considers in fact that ether is "continuously in the air, in the eyes, in the nerves, and until then in muscles"[30]. It is the environment that allows a link of dependence to be set up between a light source and the movement of muscles. Subsequently, G.-L. Buffon continued this explanation, mechanical in nature, in the relationships of every living organism with its environment, then J.B. de Lamarck (who was a pupil of G.-L. Buffon and a private tutor to his son) took it up in turn by talking about "influential circumstances". Multiple well-documented cases of physiological adaptation highlight this truism of the need for this property in order for life to be maintained.

---

30 Canguilhem, G. (1980). *La Connaissance de la vie*. Vrin, Paris, 131.

2

On the Nature of the Contribution Made by Mathematics to Biology

> *Mathematics is the language that allows a question to be expressed and resolved.*
>
> W. Heisenberg[1]

In referring to this quote, the classic approach of a physicist, we have the intention of continuing on from C. Bernard's injunction that we cited previously, since it is important to present problems correctly and not just to calculate on the basis of what is observed. On the margins of this double aspect of circularity (back and forth question–answer) that we will be illustrating throughout the various chapters, it is necessary to note an obvious and fundamental epistemological difference between mathematics and the various sciences that are known as experimental.

In mathematics, as a starting point there is always a precise set of definitions, axioms and postulates, a basis that we take care to specify thoroughly and which acts as an unambiguous label for any work done by a mathematician. The ultimate goal is a demonstration or proof, leading to the statement of a lemma or a theorem. At the very least, importance is given to demonstrating the existence of a solution that will then need to be specified. For example, the existence of a unique solution to a first-order differential equation, $y'(t) = f[t, y(t)]$, is clearly formulated by the Cauchy–Lipschitz theorem (or conditions), which constitutes a preliminary to the use of dynamic systems. Just like in a waiting position, there is the case of "conjectures", which are intuitive proposals for which a comprehensive and total

---

[1] Cited by Lévy-Leblond, J.-M. (1982). *Penser les mathématiques. Séminaire philosophie et mathématiques de l'ENS*. Le Seuil, Paris, 196.

demonstration is still being sought[2]. In biology, on the contrary (just like in other experimental disciplines), we cannot truly consider this a demonstration. We seek to *describe* and to *represent*, even to predict, using laws or models to perfect what is expressed in a first instance by the direct conclusions of a set of experimental results or observations.

2.1. On the affinity of mathematics with biology

Let us provide a few more details about this fundamental question of the nature of mathematics with regard to experimental disciplines. We know, for example, that H. Poincaré (*La Science et l'hypothèse*, Chapter 6) refuses to consider geometry as an experimental science. Although it is the branch of mathematics that is closest to a physical reality, it does not manipulate in any way the concrete and tangible objects on which it carries out its material experimentation. "The principles of geometry are not experimental facts" and no experiment, he insists, can contradict (by definition) both the postulate made by Euclid and that made by N.I. Lobatchevsky, each of which is the foundation, in an irreducible manner, of a precise type of geometry[3]. This does not prevent H. Poincaré from, of course, admitting that "experiments play an essential role in the genesis of geometry". We should not forget that, originally, geometry was a tributary of tangible problems, recalling the classic example of calculation of the height of a physically inaccessible point that is resolved by Thales' famous theorem using the shadow cast by this point. Moreover, it was associated, through drawing, with the practice of making constructions using a ruler and compass. The name "geometrical construction" thus makes a lot of sense. On this point, let us remind ourselves on the relationship that is established between projective geometry and its applications that were developed, after the work of Vitruvius, in the ancient treatises on shadows in architecture. The mathematician G. Monge, the inventor of descriptive geometry (a branch of mathematics that was taught in the past in secondary education alongside analysis or trigonometry), took a keen interest in this. It therefore should not be forgotten that mathematics, just like experimental sciences, has two faces, namely a science of observation and a science of representation[4]. An affinity is therefore presented between biology and

---

[2] We can cite the well-known case of Fermat's theorem, which although stated in the 17th Century, was not fully demonstrated until the 1990s (A. Wiles) after a series of intermediate proposals. Today, we express this theorem in the following way: the equation $x^n + y^n = z^n$ for n whole number > 2 only has a solution if one of these three numbers, namely x, y and z, is zero.

[3] Which raises the question of the appropriate choice of the type of geometry depending on the targeted objective, opening up the debate about the dilemma of conventionalism/realism, will be discussed later on.

[4] Bruter, C.P. (1996). *Comprendre les mathématiques*. Éditions Odile Jacob, Paris, 34–37.

mathematics, erasing or qualifying the inevitable but sterile opposition between tangible and abstract.

For any degree of mathematization of the model that describes a given phenomenon and of the degree of experimental validation that is associated with it, it still obviously remains the case that *modeling is not proof*. Resorting to a quibble of language, we can simply say that "such and such a model 'shows' that...". By this we mean that "the model sets out a set of observations". In other words, we establish an agreement between what we measure and what the model expresses, meaning a concordance between the physical description by observation and the mathematical description by formalized model. There is a coherence between them. On condition that experimentation is on the model itself, its simulation function contributes to providing a basis of explanation for the process studied. In other words, *the model guides or enlightens*. We reiterate its principal advantage which, except for any prediction targets, lies in bringing to light any new properties or characteristics.

The connections maintained by these two disciplines have led to undeniable successes at the interface between them, from which biology has widely benefitted, much more so than from the importance of its own contribution to mathematics. An overview of their history (Chapter 3) highlights a certain number of major points of reference in their relationship, which were at the origin of a fertile renewal of points of view by means of opening up methodological and conceptual thinking. We can state that mathematics, following the example of physics and chemistry, made a real contribution to the advance of biology. Nevertheless, these relationships still remain rather ambiguous in the eyes of numerous experimenters. Is it just slow progress that could explain the (unfavorable) comparison that is usually made with the older, and above all narrower, links between physics and mathematics, or does biology indeed present some particular aspect that affects the relationship that it can maintain with mathematics by limiting it to certain phenomena?

We are aware of just how much mathematics has been nourished by the physical world. In addition to the practical problems of everyday life, observation of given striking particularities in our perceived world has always stimulated mathematicians to spend time on them in order to describe them and understand why or how they occur. Whilst, from this point of view, biology is not in a similar position to physics, mathematics is not delayed in studying certain phenomena in living things, seeing in this the opportunity to find new problems to be solved, which is an attitude that is entirely characteristic of this discipline.

In any case, consideration of the relationships between mathematics and natural sciences remains a trivial and surprising subject. On the one hand, references to Plato and Galileo are easily brought in, as a principle that should very obviously no longer pose an existential problem regarding the position and role of mathematics, the language of nature. But, on the other hand, it cannot be forgotten that here a question arises that still poses a problem for epistemologists and scientists. On this point, the physicist P. Dirac ponders as he writes:

> "One of the fundamental characteristics of nature seems to be that the fundamental laws of physics are expressed in terms of a mathematical theory [...] we may wonder why nature is made in this way"[5].

Although this kind of consideration has little effect on the tangible mindset of biologists, we can add that, from this point of view, it appears to be required that physics and biology, without reducing the latter to the former, should go hand in hand.

Let us specify at this point what this proclaimed affinity between biology and mathematics consists of. Thus, we can make an initial statement of affinity using the correspondences that take up position between the notion of mathematical singularity and the empirically observed existence of many "points of rupture" in biology (see Figures 4.1, 4.2 and 4.3). Such is the case of the abrupt transition between the phases of accelerated and decelerated growth (point of inflection of a kinetic) or discontinuities in the allometric relationships or in action curves (threshold). On another point, there are the analogies that have been detected concerning the notions of bifurcation (qualitative change in dynamic/brutal modification from morphogenesis to sexualization of a plant apex) or homology (connection of characters or organs between different species or at different levels of the same ontogenesis). Another point to make in this is the notion of symmetry, an intuitive notion that is so familiar to morphologist biologists prior to physicists or chemists, and to which the correlative one of polarity must be added.

Numerous biologists still see mathematics primarily as a statistical tool that allows them to lay out the interpretation of their observations and measurements with some degree of rigor. Their familiarity is now fortunately well established via the nature of the conclusions of statistical tests and their language. For example, during a comparison of two treatments, it is said: "existence of a significant difference at a given risk that is chosen *a priori*"[6]. Another aim of application of statistical tools is to order and condense representation of the observations, in

---

5 Cited in Hildebrandt, S., Tromba, A. (1986). *Mathématiques et formes optimales: l'explication des structures naturelles*. Pour la science, Paris.
6 We specify that this is the first kind of risk. The reference to the second kind of risk, which results in the power of a test, is rarely explained.

particular in the case of very large files, by resorting to exploratory methods of analysis of multidimensional data. These methods have now become widespread, although sometimes a little blindly so with the trivialization of easily available software.

The position given to statistics in biology is justified by the wide variability of measurements made on living things which cannot be reduced to the quality of measurements, in part and in contrast to physical phenomena. This did not prevent G. Teissier himself from seeing the importance of a coordinated presentation of the *Lois quantitatives de la croissance*, the title of a classic book in an era (1937) where mathematics–biology relationships were still quite restricted.

This issue of a high intrinsic variability does not mean, of course, that the use of mathematics by biologists needed to be restricted just to probabilistic tools, so they have become necessary in laboratory practices. But going further implies a slightly epistemological point of view of the status of these two disciplines. Without going into detail, we need to refer to the following few remarks concerning methodology and conceptualization.

1) First, we need to recognize that the description conveyed in vernacular language continues to be relevant when quoted in biology (as in the former "natural history"). Even if, in one way or another, biology can no longer avoid mathematics (no more than chemistry and physics), it would not be able to reduce itself just to their point of view. The naturalistic mindset of observation and intuition[7] has its own virtues, just as it does irrefutable limitations. Whilst it is no competitor for the analytical and stimulatory mindset at work when faced with a computer paired up with a sophisticated instrument of this kind, it still has an irreplaceable potential for perception of phenomena. Regarding this, it is a good idea to constrain ourselves to a sort of open mind, which is, of course, valid in both directions. We can therefore use the words of the phenomenologist E. Husserl (whom we know was very taken with mathematics):

> "It is important to see [...] that exact sciences and purely descriptive sciences do indeed have a link between them, but that they can never be taken one for the other and that whatever the development of an exact science, meaning operating with ideal infrastructure, it cannot resolve the original and authorized tasks of a pure description"[8].

---

7 Keller, E.F. (1988). *L'Intuition du vivant*. Éditions Tiercé, Paris; on the discovery of transposable elements ("jumping genes") in corn by B. McClintock.
8 Husserl, E. (translated by P. Ricoeur, Gallimard, 1950). Idées directrices pour une phénoménologie et une philosophie phénologiquement pures.. In *Les Mathématiques*, texts chosen and presented by N. Chouchan, GF Flammarion, Paris, 205.

2) For its part, mathematics attaches itself in its applications to an idealization of what can be observed. It aims in the end to "replace real objects by ideal objects" (A. Lesne). Yet this transition is not automatic, because mathematics has its demands that biologists must become aware of, according to the pertinent remark made by the biophysicist P. Delattre[9]:

> "The rigor found in mathematics is not automatically transmitted to the disciplines that make use of it. Mathematics is in fact a syntax, which only retains its value when it is applied to a specific and coherent semantic. This results in the requirement to proceed to a very careful conceptual analysis before mathematization of any kind, if we wish to avoid reaching unrealistically specific proposals, or true logical plays on words"[10] (author's translation).

Whilst it must be recognized that "unfortunately the pressure of time often obliges us to carry out calculations before understanding and conceptualizing"[11], the long period of toing-and-froing between the two attitudes is quite capable of respecting this instruction of rigor as a must. Moreover, it is a mathematician who, referring to the physicist Kelvin (1879), alerts us to the harmful effects of what he calls "algebrosis", urging us not to confuse *the formula* and *the fact*[12], repeating the order made by H. Poincaré not to confuse *the symbol* and *reality*. An equation is simply a representation of a physical fact itself, a distinction pointed out many times, for example, by the physicist W. Heisenberg, who explicitly said: "Mathematical formulae do not represent nature, but instead the knowledge that we have of it".

Under the aforementioned reservations, it must be specified that mathematics bring much more than rigor of calculation, because they can turn into an *instrument of thought*. But in what way and with what objective in mind can mathematics be a useful instrument for biologists? Let us recall its two objectives, namely to be a tool of description, of condensed representation and/or forecast, or to constitute a tool of intelligibility via the fundamental concepts that it uses. Knowing that "predicting is not explaining" according to R. Thom's famous formula, let us say that, except for

---

9 Pierre Delattre was a founder of the French Society for Theoretical Biology, in particular of its Ecoles and its annual Seminars.
10 Delattre, P. (1974). Concepts de formalisation et concepts d'exploration. *Scientia*, 109, 427–458.
11 Bailly, F., Longo, G. (2008). Situations critiques étendues: la singularité physique du vivant. In *Déterminismes et complexités*, Bourgine, P., Chavalarias, D., Cohen-Boulakia, C. (eds). La Découverte, Paris, 57.
12 Bruter, C.P. (1982). *Les Architectures du feu: considérations sur les modèles*. Flammarion, Paris, 53–72. Citing the words of the physicist Lord Kelvin (1879): "Students are simply too inclined to take the easiest path, and to consider the formula and not the fact as physical reality".

an objective of pure simulation, the mathematical model can, at the very least, be an "instrument of study", with the aim of becoming an "instrument of intelligibility" (according to S. Bachelard's expression[13]). Let us examine a few points that illustrate this major advantage of the use of mathematics.

2.2. Mathematics, an instrument of work and thought on biology

For all problems tackled, the use of mathematics leads to a condensed representation in the form of figures, graphs or diagrams that are supposed to give an image or an original view that a simple description in literary language cannot provide. Due to this, they can in a first instance be a kind of *tool for perception of the phenomenon* under consideration. Although scientists from a variety of horizons have taken time to point this out, such as the physicist and chemist R. Boyle in the 17th Century, who, going further than his work as an experimenter (refer to the Boyle–Mariotte law), sees in this a considerable advantage that he expresses in the following way:

> "Diagrams, figures, representations and models represent a considerable advantage that mathematics can provide to naturalists: they significantly help the imagination to conceive many things, and by means of this allow understanding to judge these things and to deduce new consequences from them"[14].

1) Use of this mathematics first signifies adoption of a principle of methodology that consists of clearly laying out in advance the hypotheses that are at the basis of all analysis work. Thus, the conditions that validate the conclusions will be well outlined. From this framework, it is then possible to go from a particular case to an extension that has a particular degree of generality in accordance with the objective that all science pursues as an ideal.

This is well illustrated, amongst other cases, by the conditions of use of statistics. We know that any test of statistical inference (except for non-parametrical methods) is based, no matter how robust it is, on probabilistic conditions specified *a priori*. This can be in particular the normal distribution law of the variable being studied or the homoscedasticity of the samples collected (homogeneity of variances). It is necessary as a preamble to verify the hypotheses that validate the use of an analysis method of

---

[13] Bachelard, S. (1979). Quelques aspects historiques des notions de modèle et de justification des modèles. In *Élaboration et justification des modèles. Vol. 1*. Delattre, P. Thellier, M. (eds). Maloine, Paris, 3–19.
[14] Cited in Keller, E.F. (2004). *Expliquer la vie*. Gallimard, Paris, 94.

this kind, a lack of which can lead to entirely opposite conclusions. All teachers experience this when faced with the ease of use, sometimes blindly, of "ready-to-use" software, which is practical but does not at all have the slightest rigor.

Amongst the preliminary hypotheses to any mathematical study, the question arises of the *choice of a reference frame* or a system of *ad hoc* coordinates. This fundamental question of analytical geometry is of interest, for example, in mechanics in moving from a fixed Galilean reference frame to a reference frame in movement. On the other hand, biology appears to take relatively little interest in this type of problem, although the latter arises in certain methods that it uses. For example, using the principle of all factor analysis, it is necessary to recall that the search for a latent structure of a multivaried set of observations of dimension p implies *ipso facto* a change of reference frame, going from a Euclidean reference frame R^p (space for observations), first to the orthogonal system of main axes that arises from the method (and gives an approximation of it with a dimension $m \leq p$), then possibly to a system of oblique axes that result from the interpretation of main components (according to L.L. Thurstone's idea of a simple structure developed in his work on quantitative psychometry). Another example is the representation of a plant meristem in vector analysis (caulinary or root apex) with respect to a system (known as natural) of coordinates set out by the topology of the apical dome (parabolic, for example). We see that, in these two cases, we are faced with a system known as intrinsic, separated in particular from the contingency of measurement units.

2) In addition to the preliminary hypotheses to be verified, it is worth remembering that carrying out an analysis can often require the use of certain approximations, for example a development in limited series to a given degree or even the linearization of a nonlinear system around a given singular point. It is obviously necessary to specify the conditions for validity of these simplifications.

We have a good example of this with Michaelis–Menten's model of enzyme kinetics in the simple case: 1 enzyme E – 1 binding site on substrate S:

$$S + E \underset{k_{-1}}{\overset{k_1}{\rightleftarrows}} ES \underset{k_{-2}}{\overset{k_2}{\rightleftarrows}} P + E$$

Biologists currently refer to this as a base model of substrate dependence, for example, in substrate-dependent growth models. In reality, this reference is based on the use of a simplified model that results from an approximation of the above reactional system, describing all kinetics.

This approximation consists of selecting a hypothesis of "*quasi-stationarity*", a notion that is encountered elsewhere in the analysis of other physical processes. This hypothesis states that the intermediate components of a complex reaction are formed very rapidly and that their concentration then remains approximately constant. With our example, this would be the formation of the enzyme–substrate complex for which we state $d[ES]/dt = 0$. In fact, since we are interested in the very beginning of the reaction, we thus only model the initial speed involved in the formation of the ES complex according to the well-known hyperbolic function (to saturation), the Michaelis–Menten (or Henri–Michaelis–Menten) relation, validated in experiments:

$$V_{init} = V_{max} \frac{[S]}{K_m + [S]}$$

In practice, we are therefore led to consider that these kinetics are carried out in two different phases. In a transitional phase, there is a rapid formation of the ES complex according to a reaction that is supposedly irreversible. A second phase follows this where the complex is in a state of equilibrium, summarized as the consumption of S and the production of P, and for which we allow this approximation of quasi-stationarity. The calculation demonstrates that the characteristic time of the transitory phase is completely negligible with respect to the characteristic time of the quasi-stationarity phase. In other words, the quantity of substrate consumed during the transitory phase is generally very low with respect to the initial concentration $[S]$ ($t = 0$).

In summary, other than the fact that the formulation is based on the initial speed measurements for which the quantity of substrate transformed is negligible, and that there is an excess of substrate with respect to the enzyme, the following hypotheses have been selected:

– condition of equilibrium state: we suppose that equilibrium of the formation of ES is reached very rapidly and is maintained throughout the duration of the reaction, meaning $k_2 \ll k_{-1}$ et k_1;

– condition of quasi-stationary state: the speeds of formation and decomposition of ES become very weak with respect to the speeds that affect S and P, meaning that k_2 is no longer negligible with respect to k_1 and k_{-1}.

3) A related question concerns the *structural stability* of models (see Chapter 5). By "lack of structural stability", we understand the possibility of qualitative change of dynamic following a variation of the value of a parameter. In particular, we

recognize the threshold effects that allow us to move, through a minimal change of a parameter in a given region of values, from one type of dynamic to another, from which an uncertainty can arise, both in the explanation and in the prediction. In fact, biomathematicians are steered towards carrying out experimentation on the model itself (*in silico* or "computational" biology).

4) Other than these considerations of a practical nature, we need to emphasize this essential point of accepting that mathematics provides its users with fundamental concepts that are likely to direct or renew problematics. In a first instance, it is a case of the importance that needs to be attributed, as indicated previously, to the existence of singularities during a given process, a valuable indication for the human spirit which is always seeking the occurrence of something perceptibly discrete or discontinuous within a continuous environment that appears to be without form or at least difficult to apprehend (a little like a form with respect to the substrate that generates it). This poses the problem of the stability of the singularities that are brought to light, leading us to the related notion of an invariant as an intrinsic characteristic of a structure or of a process.

REMARK.– The term "invariant" is used in biology in the particular context of morphological invariants (organizational plans of the body) or chemical invariants (macromolecules, DNA and RNA, basic metabolic cycles, genetic code).

In a general manner, we recall alongside J. Monod that "the fundamental strategy of sciences in the analysis of phenomena is the discovery of invariants"[15].

Correlatively, we have three connected notions that are completely essential, namely the existence of *multistationarity*, of *bifurcation* (qualitative change in properties) and of *attractors*. Fundamental in themselves for mathematicians, they are also fundamental for biologists in the study of numerous processes. Thus, all ontogenesis goes through a series of phases that signify a change of "strategy" with respect to the current state of the system or of its environment. For example, in higher plants, the sexualization of a caulinary meristem can be viewed as a morphogenetic bifurcation. This term bifurcation, with both mathematical and biological connotations, is associated with the relatively general idea of an epigenetic landscape, *a priori* with several possibilities. We pick up the principle of the notion-metaphor of "chreode", from the biologist C.H. Waddington, who instigated meetings between biologists and mathematicians with a view to setting up the foundations of a theoretical biology[16]. Therefore, there is a "choice", at a given moment and according to the initial state of the system, that will bring a qualitative

---

15 Monod, J. (1970). *Le Hasard et la nécessité*. Le Seuil, Paris, 116.
16 Waddington, C.H. (ed.) (1968). *Towards a Theoretical Biology*. Edinburgh University Press, Edinburgh.

change in behavior, a change determined by the parameters or the variables that the mathematician attempted to put into an equation and that the biologist seeks to interpret.

The occurrence of several states of equilibrium signifies the existence of several fields that each have their own dynamic properties, an essential question for biologists. An example providing a suggestion is the phenomenon of competition between several species for which resorting to an appropriate mathematical model allows a sort of typology of cases to be obtained where the competition is expressed as an exclusion or by a coexistence. When there is predation, it is again a mathematical model that can specify the conditions (speed of growth, interactions) that do or do not induce a self-maintained oscillatory behavior. With a mathematical basis of this kind, optimal control (mathematical optimization) is able to add information about the possible intervention by the experimenter to master predation or parasitism.

A well-defined situation is that of a differential formalism jointly involving components of reaction and diffusion, and which takes an interest in the spatio-temporal evolution of the system, from which we can mathematically deduce the existence of singularities and specify their conditions of occurrence. These can correspond to localized cellular differentiations, and systems of equations of this kind are entirely appropriate for representation of organized processes in time and on a given substrate, like those that play a role in morphogenesis. We will give a few details of these with A. Turing's systems generating emerging structures.

In a very general way, finally, we highlight the importance of the mathematical concept of "optimization" for biologists in its use in seeking extremums of a function. Pointing out singularities of this kind in the sequence of variations exhibited by a given biological process corresponds exactly to this well-known observation that *"physiology is a problem of minima and maxima"*[17].

Thus, for many problems, biologists are invited not to consider mathematics under the single aim of a numerical, quantitative description of reality, as if, by summarizing a little, it were enough to calculate from an *ad hoc* formula. It goes without saying that this would not be sufficient because, in the long term, *biologists expect a pertinent representation* that can help them understand living things, at any scale of observation. Implicit in the above, we can present the following hypothesis.

---

17 Murray, C.D. (1926). The Physiological Principle of Minimum Work. *Proc. Nat. Acad. Sci.*, 12, 207–214.

PROPERTY.– For each fundamental mathematical property in the representation of a given biological system, there is necessarily a corresponding biological property that is also fundamental.

Mathematics, a guarantor of order and coherence, can be a "revealer" of the characteristics of living things in their structure and its functioning.

These various remarks are capable of explaining and specifying at the same time the affirmation by the philosopher E. Kant (1724–1804) that: "in any particular theory of nature, there is only as much science as there is mathematics"[18]. Although E. Kant was referring essentially to physics, it is necessary to extend his words to biology as the physiologist C. Bernard perceptively already desired far in advance of the time when mindsets and methods were ready for this link. Chapter 5 of this book illustrates these words with various examples, revealing to us that there can be a narrow conjunction between a given mathematical tool and a given observed qualitative characteristic. Whilst waiting, we cite here one of these cases, namely the distribution of the growth activity within the apical meristem (like a bud in the higher plant). Observation under the microscope of the frequency of cell divisions and their orientation is found to be transferred by bringing to light specific mathematical properties, according to which the direction of mitoses is organized at what is known as the principal axes. These axes are mathematical entities highlighted by a specific tool that is the "growth tensor", a tool that goes far beyond the first notion of speed.

It is interesting to note that, as a counterbalance to the fear of naturalists whose familiarity with tangible things often makes them skeptical with regard to abstract ones, mathematics itself is often the result of a progressive conceptualization that contributes to a kind of experimentation. Let us return for an instant to this fact that mathematics, motivated by a constant attempt to represent and understand the world, is often "the result of a praxis"[19]. We all have in mind the case of Euclid, who laid down the first axiomatic as a foundation for elementary geometry, whose logical consistency doubled with its application in the representation of usual physical space. However, on the other hand, Greek geometry was also constructed by "a long practice of abstraction". A good example is the calculation of surface areas. Provoked by practical requirements, this calculation led to the famous notion of quadrature (seeking a square that has the same surface area as a given figure).

---

18 Kant, E. (2017). *Premiers Principes métaphysiques de la science de la nature*. Vrin, Paris.
19 Bailly, F., Longo, G. (2008). Situations critiques étendues: la singularité physique du vivant. In *Déterminismes et complexités*, Bourgine, P., Chavalarias, D., Cohen-Boulakia, C. (eds). La Découverte, Paris, 58.

Before the advent of infinitesimal calculation, this problem was illustrated in concrete terms by the search for its solution using a ruler and compass, a technique of the work of geometrists that we have already mentioned. Let us go much further by reminding ourselves of the irreducible part played by the "perception" of objects and of their variation in all mental activity. Perceiving and studying a process means recalling that "the most stable physical phenomena at our scale of perception are those that have given rise to fundamental notions"[20]. Let us add that whilst the search for an *invariant*, meaning something stable, constitutes the main objective for all analysis, we will need to specify that in its approach or in its vicinity the direction of the variations (+ or −) is also an qualitatively essential point (state related to derivations of a higher order). *It is essential to know what does not change during a process, and also the way things that change vary.*

All these considerations are applied by constructing mathematical models. Going from an abstract formalization on the basis of well-explained hypotheses to its functional implementation through calculation and graphic images is the same as allowing certain fundamental properties of the object seen in this way to be brought to light. Thus, we aim, in addition to any prediction objectives, not only to describe the essential dynamic characteristics, but also to test hypotheses on the underlying mechanisms. All this still raises two kinds of questions. An initial problem is the identification of parameters. This is something that every modeler known very well, in particular when there is multistationarity, because then the dynamic of trajectories is determined both by initial conditions and by values assigned to the parameters. On the other hand, the relevance of the work is also based on the structural stability of the model itself, meaning its relative qualitative independence with regard to a minimal change of parameter values.

Modeling is similar to simulating real life, knowing that there is always a gap or a distance between what the model constructs and what the observation offers to us. Whether we place ourselves or not in an objective of explanation, we operate what we call a *"phenomenological reconstruction"*. Two examples can be given of this with regard to the significance of modeling. Development biology gives us a first case which involves plant architecture models[21]. These lead to quite spectacular imagery of the informational construction of a variety of species, from the very diverse form of leaves or filaments to the construction of arborescent species. In this way, we can follow the progression of plant ontogenesis both as an isolated plant and as a population, with the modifications related, for example, to the population density. This imagery of plant architectures is founded on discrete algorithms, such

---

20 Bruter, C.P. (1996). *Comprendre les mathématiques*. Éditions Odile Jacob, Paris, 16.
21 For example, the software GreenLab developed for different situations and species of higher plants.

as the deterministic language of L-systems or the principle of a probabilistic functioning of meristems that generate the formation of new modular units. This example of graphic simulation, of which we know various applications, is typical of strictly phenomenological modeling.

The considerations that we have just presented about the role that mathematics can play in biology obviously need to be placed in a wider context of scientific knowledge in general. Here we bring up an old debate, always marked by strong positions about the distance between reality and its mathematization, from Plato and his world of Ideas up to L. Wittgenstein and A. Lichnerowicz, the remarkable contribution of which was to the way mathematics is taught in secondary education, and including the slightly excessive Kantian point of view advocating that this discipline can explain everything, something of which B. Pascal was highly suspicious. Let us stop here by noting some essential points that biologists and mathematicians would be wise to keep in mind in their collaborations, if only due to the frequent lack of epistemological culture in the education of students on scientific university courses. An overview of the main stages that illustrate the contribution of new paradigms or new concepts throughout the history of biology (see Chapter 3) will allow the distinction between two very different schools of thinking to emerge quite clearly, more or less abruptly but always present. Through this we understand what is known as "realism" and "constructivism". We outline the subject as best we can within this book.

The position known as "realistic" asserts the existence of an independent reality of the human mind. The world exists as it is, without a necessary link with the representations that we make of it. We talk about an ontological existence of real things, independent of the observer. This distinction between perceived reality and, let us call it, objective reality dates back to Plato. Thus, mathematical things have their own reality and are similar to Plato's Ideas. This thesis, adopted by many mathematicians, is firmly repeated by the theoretical physicist R. Penrose (1997) for whom:

> "Each time the mind sees a mathematical idea, it makes contact with the Platonic world of ideas [...]. When we 'see' a mathematical idea, our conscience penetrates into this world of ideas and makes direct contact with it".

This vision is organized around the existence of three different but connected worlds: (i) the Platonic world (the essence of mathematics, mathematical beings and their laws), (ii) the physical world (tangible reality) and (iii) the mental world (analysis and modeling that we carry out and that provide us with "images" of perceived reality). The rational explanation that we propose for a given phenomenon or a given tangible object belongs to the Platonic world. Based on this reasoning, J. Ricard gives the examples of the trivial notions of communication and organization that biology now significantly makes use of at different scales of perception of living things. We do indeed see that these fundamental notions can only be expressed in logical terms using the idea of probability, where this is coupled with entirely abstract axioms[22]. This is thus a declaration of the degree to which the perceived tangible world requires the abstract world to define and explain it.

On the other hand, the position known as "constructivist" is based on an entirely different principle, which states that mathematical things are only idealizations proposed by us and pertaining to objects from the perceived world. For M. Planck, for example, "the world that surrounds us is in its entirety nothing more than the totality of the experiences that we have of it"[23]. For his part, G. Bachelard asserts: "Nothing is self-evident. Nothing is a given. Everything is constructed"[24]. In reality, this name "constructivism" encompasses a variety of attitudes, sometimes contradictory. Thus, for H. Poincaré[25], "scientific facts are just raw facts expressed in a comfortable language". Referring to this term "comfortable", with the connotation of a practical choice, or by convention, this position is qualified as "comfortism" (in French "*commodisme*") or "conventionalism". It demonstrates the pragmatic, non-doctrinal nature of H. Poincaré's reflections. By admitting that language and scientific principles are conventional in nature (not totally arbitrary and involving no notion of a whim, he specifies[26]), he shows a point of view, shared notably by P. Duhem, that clearly distinguishes him from certain epistemologists, in particular from G. Bachelard[27].

---

22 Ricard, J. (2008). *Pourquoi le tout est plus que la somme de ses parties: pour une approche scientifique de l'émergence*. Hermann, Paris, 248.
23 Planck, M. (1963). *L'Image du monde dans la physique contemporaine*. Éditions Gonthier, Paris.
24 Bachelard, G. (1970). *La Formation de l'esprit scientifique*. Vrin, Paris.
25 Poincaré, H. (1911). *La Valeur de la science*. Flammarion, Paris.
26 Poincaré, H. (1968). *La Science et l'Hypothèse*. Flammarion, Paris, 151–153.
27 G. Bachelard clearly expresses his opposition to H. Poincaré's conventionalism, saying in particular: "When Poincaré demonstrated in the past the logical equivalence of the various geometries, he confirmed that Euclid's geometry would always remain the most comfortable and that in the event of conflict of this geometry with physical experience it would always be preferable to modify theoretical physics than to change elementary geometry." (Bachelard, G. (1971). *Le Nouvel Esprit scientifique*. PUF, Paris, 40). But G. Bachelard goes further by talking about a "geometric subconscious" whose effect is to "immobilize the perspective of intellectual clarity" (p. 41)! *A contrario*, we know of the arguments presented by H. Poincaré on his conception of space: "I am not splitting geometry from experiments. [...] experiments on solids have just been an occasion that, amongst all the continuous groups for which we could have established a geometry, has made us choose the Euclidean group, not as the only true one, but as the most comfortable" (Poincaré, H. (1968). *La Science et l'Hypothèse*.

Over and above these epistemological distinctions (which are not at all secondary, but that we are not able to look at in further depth here), let us say that "objective reality does not consist of the content, but of the structure and the relations". Science cannot reach the things themselves, but simply "the relationships between things; outside these relationships, no reality can be known"[28]. Thus, it is up to mathematical laws to express what the human mind considers to be the harmony of the world. In one sense, this refers us to Aristotle's notion of form (*lato sensu*).

Flammarion, Paris). This pragmatic side distinguishes him in particular from Riemann, adept at a "pure", abstract geometry, that conveys to him a wider influence. On this subject, let us recall that Riemann geometry (introducing the notion of curvature of space) was admitted by A. Einstein as a structure of the universe, allowing him to go further than H. Poincaré and to establish the theory of general relativity, using tensors as a tool. This geometry is more general than Euclidean geometry and encounters the latter as a particular case by cancelation of the curvature tensor. Here, we have a typical case of a general formalization that encompasses simpler individual formalisms. Away from the essence of this debate, we can observe a significant divergence that is psychological in nature. Effectively, H. Poincaré's reflections are not restricted to his outward conventionalism, rebuffed by G. Bachelard, but are characterized by a great subtility of mathematics that meant that he embraced various aspects or correlates of the question that he had in front of him (e.g. its reference to S. Lie's groups of continuous transformations), a subtlety that does not always appear to be well understood. Therefore, we observe that H. Poincaré's position is indicated by a relatively uncommon flexibility, that it is interesting to encounter again in C. Bernard concerning vitalism, a concept that the latter rejects as such, whilst recognizing the relevance of some of its aspects (refer to section 1.1.1). Refer to the article by Michel, A. (2004). Poincaré et la théorie de la connaissance. *Philosophiques*, 31(1), 89–114.
28 Poincaré, H. (1968). *La Science et l'Hypothèse*. Flammarion, Paris, 25.

3

Some Historical Reference Points: Biology Fashioned by Mathematics

This chapter aims to give details about a certain number of founding references that laid down the basic principles that have renewed and enriched the quantitative approaches to the study of life. Concerning this term – a "quantitative" approach – we note in passing the diversity of the terminology. The term "biomathematics" automatically succeeds the former name, sometimes a little vague, of "quantitative biology"[1]. However, the term "biometrics", although from the same etymology, is confined by its use to probabilistic and statistical approaches only.

An initial insight into the place of mathematics in biology is provided in the notion of *symmetry*. We mention this briefly. In fact, the idea of symmetry took root in biology very early on, associated with the prevalence of certain numbers that are frequently encountered in the description of a morphology, considered here from a static point of view, and specifically in the case of plants. Other than the highly remarkable case of phyllotaxis (or phyllotaxy), which we will return to later on (with its dynamic aspects), the diagnosis and classical classification of species have always used characters of morphological symmetry. Thus, the arrangement of branching on a plant axis (alternate, opposite, whorled arrangements depending on the number 1, 2 or *n* of lateral appendices in each node) is a basic characteristic that participates in the facies and the habit of plants, with significant physiological implications (perception of photosynthetic radiation). We can also recall the interest attached to the *floral diagram* or projection onto a plane of the various floral parts that constitute this kind of symmetry, which we express with the *floral formula* (number and geometrical arrangement of the various floral whorls). The first rudiments of plant morphology indicate in this way various elementary types of

---

1 For example, the well-known *Cold Spring Harbor Symposia on Quantitative Biology*, held annually in the United States since 1933.

floral system, radial (actinomorphy) or bilateral (zygomorphy), and the prevalence of symmetries of order 3 (monocotyledons) and of order 5 (dicotyledons), with their variations in terms of multiplication of the number of whorls or in their movement. For example, the type 3 that becomes $2 \times 3 = 6$ for certain families (iridacaes, liliaceaes) or the coexistence of type 4 (sepals and petals) and of type 6 (stamens) (cruciferae). In addition, a polarity is added to this, of which the importance in animal embryogenesis is known, as well as in all plant ontogenesis where, from the time of germination, the development is determined by the two opposite poles, racinary and caulinary, or even the apical growth of mycelial or algal filaments.

Let us round off these preliminary remarks by stating the importance of symmetry at a molecular level (question studied mathematically), in particular the chirality of certain biological molecules with two enantiomeric forms (symmetrical figures, but which cannot be superimposed, like the image given by a mirror). For example, the forms d and l (differentiated by their optical activity in polarized light, known as dextrogyre or levogyre) amino acids or certain sugars ("oses") – glucose, fructose, galactose, lactose, maltose). In their natural state, these molecules are always of a particular type, such as natural amino acids that are of type l, with a few rare exceptions. Better still, the physiological function can be conditioned in certain cases by the type of symmetry, where only one of the forms can be assimilated and ensure a normal metabolism, and where the other can turn out to be inactive or even pathological.

We will begin by detailing two essential stages that have had an effect on the type of mathematical formalism, respectively the continuous formalism dating back to Antiquity, then the discrete formalism from the Middle Ages, two approaches that still remain in competition today. Then, with the advent of "classical science", we will observe the emergence during the Renaissance period of the notion of a law, precursor of a mathematical model. We will then review some key points that punctuated the progressive development of connections between biology and mathematics from the 19th Century onwards, up to the constitution of what is known today as the methodological corpus of biomathematics. At that point, we will no longer follow a strictly historical order due to the sometimes highly interconnected nature of the key points. We will instead seek to highlight how the different types of approaches can be opposed or mutually enriching.

This presentation aims to underline the essential points that are in our view the most pertinent in order to position the development of biology that we could describe as "formalized", remembering that this term means the revelation of suitably "mathematized" properties or relationships. The psychological and/or sociological aspects and constraints that relate to the success or the fecundity of some of these ideas will therefore be set to one side, since some of these do not yet have a place in reviews of the "state of the art" of biomathematical modeling. Similarly, certain epistemological points will often only be evoked, underlying in particular the use of the terms "theory" or "axiomatic", terms for which justification is ongoing in biology[2].

3.1. The first remarkable steps in biomathematics

3.1.1. *On the continuous in biology*

Aristotle, by basing his representation of the world not on numbers but instead on **the continuous**, stands out from his predecessors such as Pythagoras and Plato[3]. This is because Aristotle's "Physics" (in the sense of "natural sciences") emphasizes the notion of **change**, deemed essential. In particular, all living things are subject to "transformations" (= Greek "*metabolê*"), whether they are processes of growth, decline, alteration or generation. The importance of this idea continues to be emphasized, taken up again, for example, currently by F. Jacob, underlining that "an organism is never just a transition, a stage between what was and what will be"[4].

Concerning this pre-eminence attributed to change, we can observe in passing that Aristotle's thinking is both closely aligned and contradictory to that of his predecessor Heraclitus. For the latter, the dominant factor (at least for our subject matter) is the existence of a "continuous flow" in the Universe, according to some of his expressions that have become famous, like "the sun is new each day", "no man ever steps in the same river twice". The importance of this general flow is so great that fundamentally, there is precedence of the movement on Beings. Aristotle does not do this, but instead maintains their duality under the form that we will see later

---

2 For example, the presentation of the history and epistemology of these issues in the enquiry carried out by F. Varenne (Varenne, F. (2010). *Formaliser le vivant: lois, théories, modèles ?*. Hermann, Paris) can be consulted.

3 Concerning "Platonic biology", we will only mention here the essential points that have some relation to our study, as a reminder: (i) the concept of a living thing as the association of a body (made up of the four elements, earth, water, air and fire) and an immortal soul, which moves from body to body (metempsychosis), in other words, with a finalistic objective; (ii) the association of life and movement (only the soul undergoes a circular movement known as a "circuit"). While for Plato, nature itself as a whole is seen as a living thing, no reference seems to be made to plants. Moreover, his famous allegory of the cave is a general interpretation of the knowledge of the perceptive world: the only part of the outside world that the man chained up in a cave opposite the light outside can see is the shadows projected onto the wall, meaning he is without access to the exact nature of things. Concerning Plato's thinking postulating the existence of a world of ideas, see Chapter 2.

4 Jacob, F. (1970). *La Logique du vivant*. Gallimard, Paris, 10.

on. However, Heracliteus' reflections are in agreement with Aristotle when he also asserts: "All things that are in opposition cooperate. [...] All things come into being by opposition".

Aristotle's position is entirely fundamental, as pointed out by, for example, both the mathematician R. Thom (*Esquisse d'une sémiophysique*, 1988) and the doctor and philosopher G. Canguilhem. We know how much Thom insisted on the continuous nature of the phenomena and set himself the objective of *"tracing apparent discontinuities back to the manifestation of a slow, underlying evolution"*[5].

We know that Aristotle was very interested in biology, dedicating several books to it such as *History of Animals*, *Parts of Animals* and *Generation of Animals*. His point of view on this subject is entirely remarkable, since it is associated with detailed research into a variety of living organisms (except plants, however) by means of fine dissection[6] to discover their structure, accompanied by philosophical developments of his thoughts relating to various fields (e.g. his concept of the principle of causality). For the subject we have in hand, we can recall one of the principal characteristics of his "method" that he explains in the following way: "I want to talk about the question of knowing whether each being needs to be considered separately and defined in isolation, [...] by looking at them individually, or whether it is firstly necessary to carry out general research into the characteristics that are common to all of these animals". Let us consider this a premise of the 19th-Century definition of biology, when, having gone beyond the description/classification stage, the study of the two aspects, namely the unity and the diversity of living things, were assigned to it as a fundamental characteristic.

In his famous treaty *On the Soul*, he lays out his definition of life as the autonomous accomplishment of a particular potential that maintains conformity of its main themes. This notion of autonomy highlighted in this way is explained with its theory of hylomorphism which postulates that every being arises from two principles: (i) matter (*hyle* = wood in the sense of a construction material); and (ii) form (*morphe* = figure).

Matter is simply the substrate of form that is in the process of being acquired, which means that the dynamic of the act of growth conserves **form** (meaning the formational principle) and not matter. The distinction between the act accomplished and the remaining potential is added to this basic duality, i.e. between the current state (what has been carried out) and the distance with respect to a characteristic

---

5 Thom, R. (1991). *Prédire n'est pas expliquer*. Eshel, Paris, 62 *sq.*
6 We can recall, for example, that his name remains associated with the anatomical structure known as "Aristotle's lantern" (masticatory apparatus of a sea urchin).

limit value (that which remains to be acquired). Aristotle therefore postulates the existence of an *ago-antagonistic couple* as a dynamic principle: "existence in the act" (entelechy) and "existence in power". The importance of this couple was later resumed in the constitution of the foundation for certain mathematical models, in particular the well-known logistic law of growth.

It was not possible for Aristotle to represent these concepts as equations, if only due to the state of advancement of mathematics in his era, which was governed by a metrical concept of space using Euclidean geometry. Yet, for Aristotle, *a biological form cannot be reduced to a geometrical form since life means change*, something that we express today by stating that "life is a process". In fact, regulation functions have demonstrated to us since the time of C. Bernard that what matters is not so much the distances between the parties as their relationships. Canguilhem summarizes this point of view with these emblematic words: "The whole is at all times present in each part"[7].

REMARK.– This dualistic idea is an underlying basis for the mechanical concept of movement as defined by Aristotle. For example, he divides the trajectory of a projectile into two distinct parts, outlining the existence of a transition, a characteristic change from one to the other. We know that this interpretation was rejected by Galileo, who postulated the existence of two forces, one due to the initial launch and the other one due to the earth's gravitational force. This results in the famous parabolic trajectory of falling bodies.

3.1.2. *On the discrete in biology*

Well before the introduction of the notion of a determinant (Cardan, 16th Century) and the development of matrix calculation in the 19th Century, the principle of discrete formalism appeared in Medieval times with L. Fibonacci (also known as Leonardo of Pisa) (1175–c.1250). His book *Liber Abaci* ("The Book of Calculation" or "The Book of the Abacus", 1202) contains the famous series of numbers known as the "Fibonacci sequence" which is specifically presented for the description of biological growth in numbers.

The Fibonacci sequence is defined by the following recurrence: every element is the sum of its two immediate predecessors:

$$u_n = u_{n-1} + u_{n-2}$$

---

7 Canguilhem, G. (1983). *Études d'histoire et de philosophie des sciences*. Vrin, Paris, 362–364.

which produces: 1, 1, 2, 3, 5, 8, 13, 21, 34...

The "growth rate" estimated by the relationship between two consecutive numbers has the remarkable property of converging towards the famous golden ratio (*divina proportione*):

$$u_{n+1}/n_n \rightarrow \Phi = (1+\sqrt{5})/2 \approx 1.618$$

where this golden ratio is the solution to the equation $x^2 - x - 1 = 0$ whose two roots are $(1 \pm \sqrt{5})/2$.

In its operational form (Binet equation, 1843), we have:

$$u_n = \frac{\Phi^n}{\sqrt{5}} + \frac{(1/\Phi)^n}{\sqrt{5}}$$

Geometrically, the *divina proportione* corresponds to the ratio that characterizes the segment below:

$$a/(a+b) = b/a$$

in other words, the relationship of a part to the whole reproduces the relationship of the parts between themselves.

This recurring equation is of interest to biology. Moreover, it was originally presented as a description of the growth of a population of rabbits, as shown in Figure 3.1. We point out that this outline representation uses a unit made up of the parent couple, based on strong hypotheses (maturing time, fecundity, ratio of sexes). This, of course, puts the biological scope of this representation into perspective, without losing sight of the use of the Fibonacci sequence for the description of various phenomena.

The principle of this recurrence was then generalized as follows:

$$n_t = n_{t-m} + n_{t-n} \ ; \ m \neq n$$

More recent and more detailed applications followed, such as the number of cells in filamentous organisms (algae, for example), where m and n are the lifetimes (maturing) of the two daughter cells of any mitosis.

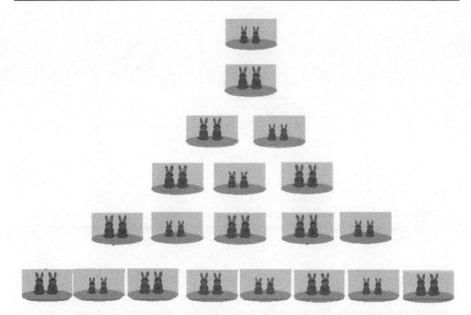

Figure 3.1. *Fibonacci sequence applied to the progeny of a couple of rabbits*[8]

Associated with this number Φ, we define the golden angle:

$$2\pi / \Phi \simeq 137.5$$

a classic reference measure in phyllotaxis that we will examine in Chapter 5. The geometrical properties of the golden ratio are well known. We cite, for example, the "golden spiral", designated as such due to its construction using the geometry of golden rectangles (rectangles whose sides produce a ratio equal to Φ) (Figure 3.2).

Corresponding to an exponential based on the golden ratio, this is simply a particular case in the family of logarithmic spirals[9]. Here, we are interested in specifying the ratio that can exist between the golden ratio and the mathematized representation of certain biological forms that feature it. This question calls for the following remarks.

---

8 Source: diagram published on the website: http://images.math.cnrs.fr.
9 The golden spiral is generally defined by the polar equation $r = a\Phi^{2\theta/\pi}$.

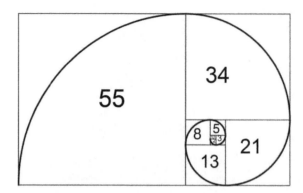

Figure 3.2. *Golden spiral constructed from tangential quarter circles in a series of squares which are themselves placed in a series of golden rectangles. The numbers (surface areas) correspond to the Fibonacci sequence*

Indeed, we are well aware of the renown of the golden ratio that we have observed to be present in various morphologies, whether in artistic, architectural or pictorial constructions (shown in particular by the writer and Romanian art critic M. Ghyka, 1931) or various natural structures, animals or vegetables. Although certain authors such as J.-P. Delahaye have criticized the enthusiasm, deeming it as disproportionate and that, in their view, plays a role in establishing a kind of myth associated with the golden ratio; focus is transferred very naturally to certain remarkable occurrences that it is difficult to describe as pure chance. The biochemist J. Yon-Kahn, among others, points this out concerning the geometric configuration of certain forms of DNA. Figure 3.3 shows measurements close to the golden ratio 1.618[10] for B-form DNA (which is the most common DNA).

Returning to our field, let us say that it is appropriate to differentiate, among natural structures that appear to develop in the form of a logarithmic spiral, between those that correspond to the construction shown in Figure 3.2, the case known as the golden spiral which is a particular case of logarithmic spirals. Analogous spirals can indeed result from the generic equation $r = a \exp(m\theta)$, i.e. without referring to $\Phi = (1+\sqrt{5})/2 \approx 1.618$. This is observed, for example, in certain systematic measurements that use museum collections containing many samples of *Nautilus pompilius*[11] mollusk shells. This data shows that the exponential argument is of the order of 1.3, therefore quite far from the golden ratio 1.618. While the conclusion drawn from this type of reading thus points out the inadequacy of the golden ratio as a basis for a logarithmic spiral equation, it does not by any means disprove the

---

10 Yon-Kahn, J. (2010). *Rencontre de la science et de l'art. L'architecture moléculaire du vivant*. EDP Sciences, Les Ulis.
11 Falbo, C. (2005). The Golden Ratio – a Contrary Viewpoint. *College Math. J.*, 36, 123–134.

principle of ontogenetic deployment of a spiral curve from this family (see Figure 3.6). *A contrario*, the real presence of the golden ratio deduced from the Fibonacci sequence is clearly testified to by recent studies on phyllotaxis in plants (this point will be detailed at the beginning of Chapter 5). We consider that the subject will continue to arouse interest from morphologists and mathematicians.

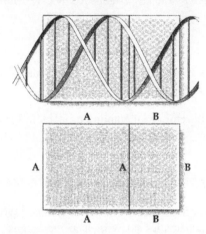

Figure 3.3. *The golden ratio and the structure of B-form DNA (A + B)/A = 1.60; A/B = 1.54*[12]

3.1.3. *The notion of laws in biology*

The beginning of the classical era is marked by the formulation of various macroscopic physical laws such as the laws of falling bodies (by Galileo in 1602), the movement of the planets (by Kepler in 1609 and 1618), optical refraction (by Snell–Descartes in 1637), and even gas pressure (by Boyle–Mariotte in 1662 and 1676), all of which are laws that are today learnt in secondary education as a foundation, and which illustrate the application of mathematics in natural sciences.

In biology, nothing like this was yet in place. However, in the Renaissance, the morphology of living things began to take shape as the subject of specific quantitative research that was admissible as empirical laws, such as the various research works and famous sketches by Leonardo da Vinci (1452–1519). Among the different morphometric questions that he was interested in, we cite the remarkable example of his law of branching in plants. The aim of this law is to numerically express the relationship between the radiuses of branches of successive order, i.e. the case of a bifurcation at node *i* of a stem:

---

12 Yon-Kahn, J. (2010) (see Chapter 3, footnote no. 9), according to Harel, R. *et al.* (1986). Beauty is in the genes of the beholder. *TIBS*, 11(4), 155–156.

$$r_i^2 = r_{i+1,1}^2 + r_{i+1,2}^2$$

This simple statistical relationship does not take into account the branching angles, nor especially the mechanical aspects of volumetric flows and viscosity[13]. It was revised and corrected by the law established by C.D. Murray (1926):

$$r_i^3 = r_{i+1,1}^3 + r_{i+1,2}^3$$

We know of various applications of this, mainly on the facies of plant branching and in the arborization of blood vessels.

In Chapter 4, we will return to the notion of a law in biology, in addition to the notion of a model.

3.1.4. *The beginning of classical science: Descartes and Pascal*

We are aware of the often-repeated words of Galileo (1564–1642) asserting that the "grand book [of mathematics is] written in the language of mathematics", in such a way that without knowledge of this language "it is humanly impossible to understand a single word of it"[14]. But at that time, it was only a question of physical objects and phenomena. Living things were not yet seen by the light of their possible relationships with mathematics. R. Descartes (1506–1650) himself hardly dedicated any time to biology, apart from his concept of "animal-machines", an idea revised and spread later by J.O. de La Mettrie (1709–1751) with *L'Homme Machine* (1748). In reality, de La Mettrie distanced himself from Descartes, the latter considering that humans are made up of a body and a soul.

In relation to this era, which was a major historical turning point, it must be stated that the point of view of Descartes is more often remembered (described as analytical and reductionist) than that, entirely different, of his contemporary B. Pascal. It is also useful to mention, on the contrary, the very important words of the latter, clearly systemic in it inspiration.

---

13 On the subject of the branching angles, we sometimes cite this type of law that Leonardo da Vinci is believed to have predicted: "the smaller the diameter of the branch, the further it is from the trunk with an angle close to 90°" (according to Blaise, F. (1991). Simulation du parallélisme dans la croissance des plantes et applications. PhD thesis, University of Strasbourg, quoted in Varenne, F., *op. cit.*, p. 108). This leaves doubts concerning the well-known plant facies known as plagiotropy (obliqueness of a branch close to the horizontal, variable depending on the species and the stage of development).

14 Galileo, G. (1623). *The Assayer*. Translated by Stillman Drake [Online]. Available at: https://web.stanford.edu/~jsabol/certainty/readings/Galileo-Assayer.pdf.

PROPERTY.– "Since all things have causes and cause in turn, are helped and help in turn, are mediate and immediate, and since all things maintain each other by a natural and imperceptive connection that connects the furthest away and the most dissimilar, I consider it impossible to understand the parts without understanding the whole, nor can I understand the whole without understanding the parts separately" (Sellier 1976, p. 132).

These lines foretell in spirit the basic principle of what is known today as "systemic biology". But, in fact, this was the point of view held by Descartes, who dominated and for a long time influenced all scientific methodology with his concept of direct unidirectional causality that results from his desire to divide the whole into elementary parts that are more easily accessible.

3.1.5. *Buffon and hesitations relating to the utility of mathematics in biology*

A recurring subject, G.-L. Buffon (1707–1788), provides a good illustration of this. The case is indeed remarkable and deserves some discussion due to his paradoxical point of view about the relationships between biology and mathematics.

The author of the monumental *Natural History*[15], Buffon takes an interest in a wide variety of subjects, such as the formation of the planets or the propagation of heat. For the latter, for example, he carried out tests on the speed of cooling of different sizes of iron balls in his smithy in Montbard, noting the ratio between the diameter and the time taken for cooling. However, before dedicating himself to his principal work as a naturalist, particularly in zoology, his primary center of interest was mathematics. Thus, he took on the translation of *Method of Fluxions* by I. Newton (1740), the origin of infinitesimal calculation[16]. He set up ongoing relations with certain mathematicians, like the correspondence that he engaged in with the Swiss G. Cramer[17] or the support he found in the academic A.C. Clairaut. He also translated the *Statical Essays* by S. Hales (1735), which focused particularly on the speed of growth of stems and leaves. On this subject of biomathematics, he wrote a dissertation on the number and thickness of ligneous layers (rings) (1737), and about the mechanical resistance of wood.

---

15 36 volumes released during his lifetime from 1749 to 1789, supplemented by eight volumes released after his death by B.G. de Lacépède.
16 *Fluxion*: derivative with respect to time, denoted $\dot{x}$.
17 On the subject of the "St. Petersburg paradox" (mathematical expectation vs. behavior of the player that restricts their participations in the game), a question of probability was debated by N. Bernoulli and G. Cramer during their era. Buffon talks about a degree of "physical certainty" (probability) as opposed to the "moral certainty" that determines the player's decision.

We know above all about Buffon's interest in the calculation of probabilities regarding the "game of franc-carreau" (fair-square game) and more particularly the associated problem known as "Buffon's needle". He enjoyed the support of the mathematician Clairaut who was a reviewer when his *Mémoire sur le jeu du Franc Carreau* was presented to the Academy of Science. Written in 1733, this report was not published until 1777 in his *Essays on Moral Arithmetic*.

The game of franc-carreau (fair-square) consists of throwing a coin onto a tiled floor, and looking at the place where it falls. The throw produces a win if the coin falls onto a tile without touching its edges (we mean "a true square"). The originality of Buffon was to consider this probability problem by relying on infinitesimal calculation.

Buffon's needle problem (Figure 3.4) considers a similar question with a needle thrown onto a wooden floor made up of wood strips of equal width l: what is the probability that a needle of length a falls onto a crack in the wooden floor? Buffon provided a theoretical expression for this:

$p = 2a/\pi b$

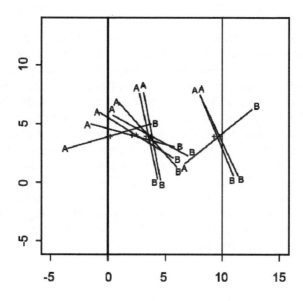

Figure 3.4. *Buffon's needle problem*

The result of a series of throws provides us with an estimate of the number π. We note the experimental validation given to this by M. Wolf in Zurich in 1850. He obtained 2,532 intersections for 5,000 tests (l = 45 mm; a = 36 mm), providing the estimate π = 3.1596 (very slow convergence).

3.1.5.1. *On Buffon's epistemological position*

Two things can explain the point of view of Buffon with regard to the possibility of a productive connection between biology and mathematics, between what can be observed and what is abstract and could describe the former. Thus, in terms of the calculation of probabilities, Buffon considers that the statement that the instance of winning a game can only increase with the number of attempted matches is in fact not realistic in practice. Faced with the theoretical concept of mathematical probability, he sees the requirement to introduce the idea of a principle of utility. We can see this association of mathematics and psychology as distantly related to the current theory of games, which is based on the idea of strategy. Moreover, Buffon mentions analogy a lot more than causality. We see that it was not possible for this point of view to favor pursual of what could be a deterministic law of nature in biology, which would be distinct from simple statistical laws of occurrence.

The case of Buffon is an exemplary illustration of what was (and sometimes remains) the school of thought of naturalists and experimenters. This is because, despite his confirmed enthusiasm for mathematics, Buffon still wonders about the existence of ambiguities in the relationships between this and biology. Better still, he points out that applications of mathematics to natural objects that are too complicated can lead to errors. The reason for this, he says, is that we are led to "remove most of the qualities from the subject, to make it into an abstract being that no longer resembles the real being". In summary, "the most delicate point and the most important [...] is to know how to distinguish correctly what is real in a subject, from what we put it in that is arbitrary [...], to clearly recognize the properties that belong to it and those that we attribute to it" (Buffon 2017). This suspicion goes a lot further than his reflections about the calculation of probabilities in practicing a game. It leads him to the conclusion that we cannot consider mathematics as something that can make a useful contribution to the description of living things.

With extremely modest results, the relationship that Buffon maintained with mathematics has nevertheless left a lasting effect. Having been the first to be closely interested in a question of *geometric probability*, his work on the needle problem is not just a diversion. Indeed, he pronounces the application of calculation of probabilities to the study of geometrical figures, therefore gaining the interest of certain sectors of biology, such as morphometry of cellular structures. This is demonstrated by the international colloquium held in Paris in 1977 for the bicentenary of the release of his *Essays on Moral Arithmetic*, and entitled

Geometrical Probability and Biological Structures: Buffon's 200th Anniversary[18]. We note among the work presented, applications of the methods of mathematical morphology to solid biological problems, both in cytology (structure of the chromatin) and in histology (tessellation or cellular paving).

Concerning the hesitations surrounding the legitimacy of the use of mathematics in biology, we know they still exist, if only due to our lack of knowledge of the issue. We have an example of this with the botanist L. Plantefol, who proposed in 1946 an explanation of the "phyllotaxy" (ordered arrangement of successive leaves along a stem) and refused to acknowledge the advantage presented by a mathematical description of the observations (see Chapter 5).

Consequently, in the 19th Century, the biology-mathematics connection began to take shape, far removed from Buffon's apprehension. Without associating ourselves with a strict historical order due to the interconnection of so many essential stages, we are going to instead consider some of the remarkable points in this development according to the type of phenomena in question (from demographics to genetics, and from morphogenesis to physiology) and depending on the imbrication of approaches.

3.2. Some pertinent contributions from mathematics in the modern era

3.2.1. *The laws of growth*

The first formulations of the growth of populations came to light in the 18th Century with L. Euler (1707–1783) who, with this in mind, noted the interest presented by the exponential function (or geometrical):

$$dx/dt = rx; \quad r = Cte; \quad x = x_0 \exp(rt)$$

as a law of growth defined by the constant nature of the specific speed of growth:

$$(1/x)\,dx/dt = Cte$$

For the economist and demographer T. Malthus (1766–1834), whose name is often associated with this formulation, this law of numbers in a geometrical progression is in balance with the arithmetical progression of resources, but a

---

18 Miles, R.E., Serra, J. (eds) (1978). *Geometrical Probability and Biological Structures: Buffon's 200th Anniversary*. Springer, Berlin. This contains an extract from Buffon's essay. See the communication by J. Roger, Buffon and Mathematics, 29–35.

general law that combines these two components is, however, not established. This was, however, already proposed in 1725 by the mathematician A. de Moivre based on the principle of a "force of mortality", which he presumed to be proportional to the remaining number of years, where this limit was then fixed *a priori*.

The first function of growth duly explained and validated was established in answer to a practical objective of a forecast of the production of wood. This is the law of Hossfeld (1822):

$$\frac{dx}{dt} = C x^{1-m} \left(1 - \frac{x}{K}\right)^{1+m}$$

always used in forestry, as such or with a few variations.

In the continuation of this demographic, the classic Gompertz (1825) and Verhulst or "logistics" (1838) functions were put forward in the same era, under the general name of "growth laws", and are both defined by a differential equation that expresses the variation of the specific speed or growth rate μ:

– logistics law: $\frac{dx}{dt} = a x \left(1 - \frac{x}{K}\right)$; $\mu = \frac{1}{x} \frac{dx}{dt} = a \left(1 - \frac{x}{K}\right)$ (linear decline of μ with the variable x);

– Gompertz law: $\frac{1}{\mu} \frac{d\mu}{dt} = -a$ (exponential decline of μ over time).

3.2.2. *Formal genetics*

Research into the distribution of characteristics within a population constitutes one of the rare examples where biology has true and specific laws that can be compared to the macroscopic laws of physics mentioned previously. These laws involve, in different forms, either the progeny of controlled experimental hybridizations or the evolution of a natural population that is subject to random crossings. In addition, formal genetics involves both the distribution of discrete or qualitative characteristics (such as a type of color) and continuous or quantitative variables (such as a dimension).

Let us briefly give an insight into the corresponding methodologies.

3.2.2.1. *Mendel's laws*

Mendel's laws (1822–1862) on the segregation of characteristics in the descendants from a hybridization between two homozygotic parents founded **formal genetics**. Published in 1865, they were completely forgotten before being rediscovered in 1900, simultaneously by three botanists, the Dutchman H. de Vries, the German C. Correns and the Austrian E. von Tschermak.

G. Mendel first notes in his first law the uniformity for F1 descendants (first generation after hybridization). His second law then specifies the genetic composition of F2 descendants (second generation) on the basis of certain specific hypotheses (see following paragraph). He therefore gives the following proportions: (3,1) for monohybridism and (9,3,3,1) for dihybridism.

The hypotheses made by Mendel are:

– all characteristics are specified by the existence of hypothetical entities (future genes), transmissible as such in the descendants[19];

– these entities or genetic determinants are presented in two forms: dominant/recessive;

– independence of characteristics is respected.

Of course, the choice of plant species was decisive in reaching the formulation of these laws. For this, in his Czech monastery in Brno, Mendel had a garden in which he proceeded to carry out multiple experiments. Following several series of empirical tests carried out in advance, he chose the pea *Pisum sativum* as the subject for his experiments, rejecting other species such as *Hieracium*. The latter species in fact undergoes apomixis, with non-sexual reproduction from diploid cells of the ovum and not for gametes and which therefore encourages maternal heredity.

On the contrary, Mendel worked on the segregation of independent characteristics. He could not have known, in his era, that the characteristics observed in peas themselves (smooth tegument (pea rich in starch)/wrinkled (sweet pea); yellow/green in color) were determined by the genes located on different chromosomes.

It remains that the empirical choice he made correctly underlines this general fact that all laws are relative to a certain field of application that sets out the limits of its validity.

---

19 The term "gene" was proposed by the Danish botanist W. Johannsen in 1909, by contraction of *pangene* used by de Vries, and as a replacement for various other names.

From an epistemological point of view, the situation is summarized well by E. Mayr:

> "Mendel used the hypothetico-deductive method. When we see how he has programmed all his experiences, how he has explained his method and how he has chosen his materials, it is difficult to stop ourselves from thinking that he already had a pre-determined theory in mind, and that his experiences aimed in fact to test it out"[20].

3.2.2.2. *The genetics of populations*

In contrast to Mendel who was interested in the descendants of controlled hybridizations, and a little after the rediscovery of his laws, another formalism took root in the context of the **genetics of populations**. Its objective is the study of the distribution of genotypes within a natural population that is not subject to a constraint of self-fertilization. In 1908, the **Hardy–Weinberg law** was put forward jointly by a mathematician (Hardy) and a biologist (Weinberg). It involves a panmictic population, meaning one that responds to the following hypotheses: high numbers, random crosses, absence of any selection, migration and mutation. For two alleles *A/a*, of probability p and q (of which none are unfavorable), the distribution of frequencies of the genotypes formed at the zygote stage {*AA, Aa, aa*} is:

$$\left\{ p^2, 2pq, q^2 \right\}; \; p+q = 1$$

This law results quite simply from the random meeting of gametes according to Mendel's diagram on the "chessboard of gametes". In principle, of course, it can apply later to individuals that carry these genotypes only by taking account of any differences that there may be in their survival and fecundity. Despite the strong hypotheses of panmixia, we observe that this theoretical distribution is often quite well verified in reality. It constitutes a structure of stable equilibrium, reached more or less rapidly.

Like those of Mendel, this law is a pioneering work. It was not before the 1920s that genetics of populations developed from the work of R. Fisher, J.B.S. Haldane and S. Wright. In this context, Fisher emitted what is sometimes pompously known as the "fundamental theorem of natural selection".

This "theorem" related to the variable known as the "selective" or "adaptative value" (*fitness*), referring to which he stated that the rate of increase in fitness of any organism at any time is equal to its genetic variance in fitness at that time. Let us take note that only the "additive variance" is taken into account (additivity of the

---

20 Mayr, E. (1989). *Histoire de la biologie*. Fayard, Paris, 659.

action of the different alleles) according to the standard model of decomposition of the genotypical variance[21]. On the contrary, the selective value of a genotype is equal, to the nearest coefficient, to the probability of survival of individuals that carry this genotype from their conception until they reach a reproductive age. In summary, we can say that this variable, rather than being a basic concept, is a calculation method that measures the propensity of a genotype to maintain itself in the population, i.e. the probability of an event occurring.

From an epistemological point of view, it is interesting to point out that Mendelian genetics on the descendants of controlled hybridizations, and on the distribution of characteristics in natural populations, both developed through the manipulation of an abstract entity, a gene, without consideration of its physical nature. It was only much later on that the latter was the subject of specific research using molecular biology and beginning with the discovery of the structure of DNA in 1953.

In reality, the very term "gene" is no longer a synonym of a segment of DNA (a *locus*) that is believed to be operational for a given function. It now designates only part of the functional genetic unit of transcription that is known as an "operon". The classic example is that of the lactose operon, genetic determinant of the metabolism of lactose (seen in the bacteria *Escherichia coli*, Jacob and Monod, 1961). The lactose operon effectively includes the following "genetic elements": three structure genes, two control sites (promoter and operator) and one regulatory gene. Let us add the specific case of "transposons" or "jumping genes" (mobile segments of DNA that can move on the chromosome). Another type of gene is "homeotic genes". This term designates a DNA sequence whose mutation causes an anomaly in the embryonic development with the formation of an organ in an abnormal position (e.g. in insects, appearance of legs in place of antenna, or, in plants, the formation of petals in place of stamen). For a given species, all homeotic genes have the same nucleotide sequence (referred to as homeobox).

3.2.2.3. Quantitative genetics

Another field of formal genetics relates to continuously varying characteristics or characteristics known as quantitative (such as the size of an organism). This is the subject of **quantitative genetics** which thus concerns multifactorial characteristics for which, due to this multiple determinism, Mendel's laws cannot apply. In reality, Mendel's vision also referred to these quantitative characteristics that he studied in particular in the flower color of the Spanish bean *Phaseolus multiflorus* (= *Ph. coccineus*), of which the heredity did indeed show that it was a continuously varying

---

21 Concerning this conclusion of an average selective value that diminishes as the selection goes on, see the comment made by the geneticist A. Jacquard (1977). *Concepts en génétique des populations*. Masson, Paris, 101–107.

characteristic that he could not divide up into a given series of shades of colors in the way that he could with two-state characteristics studied in *Pisum*. However, without resorting systematically to probabilistic considerations, he could not quantify what could have been the analogy of his segregation laws for discrete characteristics.

The principle of quantitative genetics is based on the distinction between several types of variances in the distribution of a characteristic. He aims to relate the variance of the observations, known as "phenotypic variance P", to his two components that may explain it – the variance known as "genetic G" (related to the genes that intervene in the determinism of the observations) and a variance known as "environmental E", which would result from the influence of the environment that quantitatively modulates the role of the genes at stake. A *standard additive model* is commonly used, laying out the additivity of these types of variances:

$$\text{var}(P) = \text{var}(G) + \text{var}(E)$$

On this point, we need to recall that in the statistical analyses of variance (see section 3.3), the fundamental hypothesis relates to the additivity of the sums of the squares of the differences from the average and not to the additivity of the variances, since the different components of the total dispersion do not have the same number of degrees of freedom. We therefore need to emphasize that the additivity of variances postulated in quantitative genetics is based on the hypothesis that the effect of the environment is the same for all genotypes. We believe that this corresponds to the verification of random environmental dispersion by classic experimental devices.

Some specific notions are defined. Therefore, we state that the phenotypic value P of an individual is the sum of its genetic value G and of an environmental term E:

$$P = \mu + G + E$$

where G and E have independent distributions. In the case of a single locus (case where the characteristic studied is determined by a single gene known as a main), we define the average effect of an allele as the centered expectation of the phenotypic value of individual carriers of this allele. The corresponding theoretical model distinguishes, within this term for genetic value G, the result of an additivity effect A of the two alleles, and of a dominance effect D, where A is denoted "genetic additive value". That is, for the genotype (i, j):

$$G_{ij} = \alpha_i + \alpha_j + \delta_{ij}$$

where the α are the respective effects of each of the alleles of the parents and δ_{ij} is the interaction.

As a result, the notion of "heritability" developed as a measure of hereditary transmission, relating the variability observed in the descendants of the characteristics to the relative part of what comes from the genetic value with respect to the environmental effects. In a wider sense, this parameter is written (Lush 1937):

$$H^2 = \text{var}(G)/\text{var}(P)$$

3.2.2.4. *The probabilistic point of view*

Besides these classic considerations of quantitative genetics, other much more elaborate contributions made by mathematics should be considered and at least mentioned. This is thus the case for research by the French mathematician G. Malécot (1911–1998) in genetics of populations, which he studied from a probabilistic point of view[22]. Differentiating himself from the strictly statistical approaches of Fisher and Wright (based on the calculation of correlations between characteristics and oriented towards the idea of a selective value or *fitness*), the originality of Malécot related to an entirely different concept, the pioneering idea of *genetic identity*. The objective was to point out the existence of genetic relations between individuals which, in a natural population, are connected in a number of ways by random fertilization. Measuring the degree of genetic connection between two individuals with alleles in common consists in principle of going back through the generations to find common ancestors that have left them with a particular allele. We see that this question arises from stochastic processes of branching or ramification (see Chapter 5).

His research led to the calculation of the parameters *ad hoc*, *coefficient of consanguinity* and *coefficient of parentage*. The latter is a characteristic for a couple of individuals to have had ascendants in common. Concerning the consanguinity of an individual, this is a piece of information that corresponds to the fact that their parents were more or less related. Mathematically, it is the probability that two homologous genes are identical. To summarize what is known as the "relationship of genetic identity", let us consider at a given locus and for a given couple of individuals, the occurrences of the four possible alleles (two per parent). Since these alleles can be different or identical through the series of ascendants, there is a certain number of typical genetic situations that exhaustively describe the possible combinations of these alleles. These situations, combinatory in nature, define the relationships of genetic identity between individuals. This notion of genetic identity was refined in France and in the United States in the decades 1940–1970, and was

---

22 Malécot, G. (1948). *Les Mathématiques de l'hérédité*. Masson, Paris.

founded on the notion that can be designated by the term *"genetic structuralism"*. M. Gillois (1964) then pointed out the existence of 15 different relationships, which were then considered in terms of the different ways of grouping them together (condensed relationships)[23].

Considered as a pioneering work on the subject, this research, although more restricted in its application than the previous model of decomposition of the phenotypical variance, has now been well-described by the international genetic community, with a justifiable position in treatises on the genetics of populations, in particular on the theme of *genetic structures of populations*.

3.2.2.5. The algebraic approach

Finally, we observe the development of an algebraic approach with the aim of detailing how hereditary transmission of characteristics within a population takes place (that we have seen drawn up by the phenomenological law of Hardy–Weinberg). The objective of this formalism[24] is to constitute a particular type of algebra, known as *"genetic algebra"*, whose properties correspond precisely to the laws of Mendelian genetics[25]. These are thus seen as the result of the existence of underlying algebraic structures. Without detailing this specific formalism and demonstrating its implications on the structure and the dynamic of the population, we note the basic idea in the simplest case of a gene with two alleles (*A* dominant, *a* recessive, with total dominance). Therefore, it is necessary to construct an algebra simulating the Mendelian behavior that expresses in classic terms the chessboard of gametes. While this table of contingency is read simply as the combination of elements (line × column intersections), we look for an algebra that can explain it with a series of operations between elements. We formulate this in our example by considering a population of individuals denoted u_1 and u_2 (corresponding to the two types of alleles, respectively dominant and recessive with total dominance) for which the law of internal composition (multiplication) is:

$$u_1^2 = u_1 \; ; \; u_1 u_2 = \frac{1}{2}u_1 + \frac{1}{2}u_2 \; ; \; u_2^2 = u_2$$

---

23 The first research by Cotterman dated back to 1940 (thesis), but it was not published until 1974. Consider, for example, Jacquard, A. (1966). Logique du calcul des coefficients d'identité entre deux individus. *Population*. 21(4), 751–776, with the presentation of various genetic situations and proposed groupings.
24 Etherington, I.M.H. (1940). Genetic Algebras. *Proc. Roy. Soc. Edinburgh*. B, 59, 242–258.
25 In the modern sense of the term, we understand by algebra the study of the properties of any set of mathematical beings in general (and not only of numbers), a set that contains well-defined laws of composition.

This law (which is not associative) is equivalent to the Mendelian behavior of genes A and a where the "product" AA gives A, aa gives a and Aa gives A and a in equal parts. On this basis, we can study the effect of a given mutation rate on the distribution of alleles.

The property of non-associativity is clearly shown in the descendance of a series of gametic encounters according to the method of successive crossings. For example, by representing A, B and C, three individuals who are likely to come together, and without taking into account here their allelic composition, there are two possible trees of encounters, depending on whether the first union is $(A \times B)$ or $(B \times C)$, showing the non-associativity: $(A \times B) \times C \neq (B \times C) \times A$.

REMARK.– In passing, we note the importance of this type of formalism in *taxonomy or biological classification*, a theme whose current importance is known in various fields (phylogenetics, for example). The algebraic approach renews the statistical methods of multidimensional data analyses (such as discriminatory factorial analysis)[26] by specifying the theoretical foundations of the representations of branched trees (algebra of classification trees)[27].

3.3. Introduction of the notion of a probabilistic model in biology

The use of a **mathematical model** as a tool in the analysis of experimental data is at the foundation of what we call *"statistical biometry"*. It began in agronomics for the study of nutrition of cultivated plants (fertilization/yield relationships), as well as for the comparison of different cultivars in the creation of varieties. Experimental agronomics played a pioneering role in perfecting experiment designs: sampling and arrangement *in situ* of the "individuals" experimented on, statistical analysis of variance, comparison of sample averages.

The *block randomization method (simple blocks)* by the British statistician R.A. Fisher, archetype of experimental designs, is founded on the following stochastic model, intended for the study of a factor A and of a repetition factor B (blocks):

$$X_{ij} = \mu + \alpha_i + \beta_j + \varepsilon_{ij}$$

---

[26] Since the classic research works by Sokal, R.R., Sneath, J.H.A. (1963). *Principles of Numerical Taxonomy*. Freeman, New York, then by Benzecri, J.P. (1976). *L'Analyse des données, I: La taxonomie*. Dunod, Paris.
[27] Parrochia, D., Neuville, P. (2013). *Towards a General Theory of Classifications*. Birkhauser, Bâle, 93–124.

We postulate that the experimental data X_{ij} (e.g. the biomass produced) is additively expressed by a deterministic parametric part $\{\mu, \alpha_i, \beta_j\}$ and a random term ε_{ij}. These parameters represent, other than a general average effect μ, the action α_i of the factor studied A on its modality i ($i = 1,..., p$), and that of β_j of the repetition factor B (block $j : j = 1,..., n$). By "block", we mean a set of "individuals" (e.g. p experimental plots that each receive one of the p modalities of the factor A).

Specific probabilistic hypotheses are associated with random variables ε_{ij} (independence, Gaussian distribution, homogeneity of variances) in such a way as to be able to correctly carry out a variance analysis. The principle of this consists of comparing the variance due to the study factor A and the variance of random nature, taking into account the part of variation due to the differences between blocks. The rule of decision (effect deemed globally significant between the various modalities of factor A) is probabilistic in nature, setting up *a priori* a given risk of error. Finally, the modalities of A that are different are sought out, two by two, and those that are not, still for the same risk of error. Let us specify that the validity of this model of simple blocks implies that there is only one type of variation between the repetitions. In the case of an *in-situ* experimentation in the field, this means that the environment only presents *a priori* a single direction of spatial heterogeneity (ground, microclimate). The blocks are then arranged perpendicular to this direction, in such a way that the variation between blocks can estimate this environmental cause of variation. Of course, in the case where there are clearly several causes of heterogeneity due to the environment, another model must be targeted, including several parameters that estimate these causes of variation.

In Great Britain, the first agronomic station was created in 1843 (Rothamsted station, still in use, grouped together with other stations), which was repeated in various other countries. France is well-equipped for this, with the network of stations in the INRA (French National Institute for Research into Agronomy), first in Versailles, then in various regions), added to by experimentations carried out by various professional institutions (such as the Technical Institute of Cereals and Forages), as well as by various industrialists. For example, the former French National Industry Office for Nitrogen (ONIA) had its own station for agronomic testing of compound fertilizer formulae in Toulouse. Analogous experimentations were carried out, among others, by the Alsace Commercial Company for Potassic Fertilizers. The agronomic stations of the INRA widened their initial field of study to various other sectors (improvement of plants, protection of crops, pedology, bioclimatology, zootechnics).

Concerning these questions of applied statistics, of which agronomy was the initial driving force, the important role played by Fisher (1890–1962) needs to be kept in mind, on the one hand, with his theoretical contribution about the principle of estimation of the maximum likelihood and the notion of information in statistics (we are referring to "Fisherian statistics"), and, on the other hand, with the methodology of variance analyses that he implements at the station at Rothamsted. The latter are intended for a comparison between controlled factors and random fluctuations. The previously mentioned elementary case of simple blocks known as "Fisher randomized" (1 study factor + 1 repetition factor) was completed with the simple Latin square (1 study factor + 2 repetition factors) and the Graeco-Latin square (1 study factor + 3 repetition factors), then with testings known as factorial (several study factors combined with the study of their interactions) and series of testings (on several stations and over several years). The industry appropriated this type of experimental methodology, in particular becoming interested in the estimation of relationships between the effect of a compound and its concentration. Moreover, let us give a reminder of the principle of polynomial regression that is noted in the introduction.

3.4. The physiology of C. Bernard (1813–1878): the call to mathematics

In his two significant works, *An Introduction to the Study of Experimental Medicine* (1865) and *Lectures on the Phenomena of Life Common to Animals and Plants* (1878), Bernard was the first to analyze the question of the pertinence of quantitative laws that can summarize the relationships between various biological functions. The philosopher H. Bergson considered that Bernard's works were to biology, what *Discourse on the Method* by Descartes was to physics. In reality, Bernard's reflection makes do with pragmatisms of a dialectic between the different, or contradictory, positions, notably combining "mechanicism" (determinism of physico-chemical laws) and "vitalism" (life is more than physics)[28], as we have previously noted (Chapter 1).

An experimenter first and foremost, as well as with an eye for epistemology, Bernard clearly set out his thoughts on the place for mathematics in physiology, as demonstrated by his words that we have used as an epigraph and where we believe we can hear echoes of Galileo. However, he specifies: "I am convinced that the general equation is impossible for the moment, since the qualitative study of phenomena must necessarily precede a quantitative study of them"[29].

---

28 A more flexible position, also richer than that, for example, of the doctor A. Carrel, who believed that "mechanicism and vitalism must be rejected in the same way as any other system" (Carrel, A. (1935). *L'Homme, cet inconnu*. Plon, Paris, 38).
29 Bernard, C. (1984). Introduction à l'étude de la médecine expérimentale. Flammarion, Paris.

In summary, Bernard shows great interest in the application of mathematics to physiology while at the same time considering their implementation as premature. For him, the degree of mathematization of a science is an image of its state of advancement. Let us add a certain reticence regarding the use of statistics.

Thus, on the subject of his observations on glycemia, Bernard questions the correctness of taking the average of measurements, because this hides the reality of the oscillations on a nychthemeron (chronobiology, with the analysis of time series, had not yet surfaced).

A key notion arises from the physiology of Bernard: the property of *autonomy* of living things. This property is demonstrated, on the one hand, with *autopoiesis* (continuous production of "oneself", elements, structures) and, on the other hand, with *homeostasis* (regulation of the physiological variations that ensure the internal environment is constant); two points that he was constantly focused on.

This property of autonomy of living things, which was already apparent in the work of Aristotle (refer to his definition further on), was picked up by P. Vendryès (1981), who gave it a specific interpretation[30]. It is no longer a case of linking this property exclusively to homeostasis of the internal environment and also, and above all according to Vendryès, of relating it to the variations in the external environment.

In a more nuanced way, we can say the acquisition of the autonomy of living things would mean that there is both prevalence of its own internal determinisms and reaction to external determinisms. Otherwise stated, physiology manages to admit a "certain compatibility between the autonomy of living things and its submission to universal laws"[31].

Another important contribution of the thoughts of Bernard, despite a way of writing that can sometimes cause confusion through its dialectics, is the term of autonomy of the parts of an organism that is repeated in contrast to their interdependence. He effectively asserts that these elements, although different and autonomous, do not play the role of simple associates and that their union expresses more than the addition of their separate parts.

With this remarkable phrase regarding non-additivity, the property of emergence is announced, which would be involved later on with the self-organization of biological systems.

---

30 Vendryès, P. (1981). *L'Autonomie du vivant*. Maloine, Paris.
31 Pichot, A. (2011). *Expliquer la vie*. Éditions Quae, Versailles, 347.

3.5. The principle of optimality in biology

One of the essential outcomes of Bernard's reflections was to highlight the stationarity of the internal environment of any living organism thanks to a continuous set of regulations. Added to this there is this specification that under this term of "constancy", there is in fact a trend that copes with transitory variations. Physiological invariants are still dynamic in nature; its values are located between given fluctuation limits (except for pathological situations).

This fundamental idea of control is combined with the banal observation that has already been mentioned that "physiology is a question of minima and maxima"[32]. It refers, for example in plants, to maximizing the flow of circulating sap or to minimizing the resistance to its transport. From a methodological point of view, biology is thus required to pay attention to what mathematics calls the "calculation of variations", whose objective is to precisely seek out the extremums of a function, or more generally of a functional (meaning the function of a function, for example the integral of a function $f(x)$). This is designated by the mathematical term "optimization", because this search for an extremum corresponds, by the nature of the chosen function, to obtaining an optimal solution in the exact sense of the term (optimum = the best). For example, minimizing a duration or a cost, maximizing a biomass.

In relation to these remarks, we see that, in quite a general manner, biology is presided over by a principle of optimality, meaning a *"principle of adequate design"* according to the words of the biotheoreticians N. Rashevsky and R. Rosen (1973), also following the work of D. Cohn (1954) concerning the optimization of the branching of blood vessels in relation to the "economy" of the heart motor. We can remark that this idea established itself very progressively in biology as in fact an underlying concept to various manifestations of living things, as much in morphology and ecology as in physiology, which means a more or less teleonomic condition, a good "economy of life". Furthermore, as if to satisfy a psychological requirement searching for coherence in its rational representation of the world, it is also essential in physics and in mechanics. For example, seeking out, in metric terms, a geodesy (the shortest path between two points) or, in terms of time, of a brachistochrone (the shortest time period), as well as the famous principle of least action (or of minimal work) by P.L.M. de Maupertuis, the subject of many debates with P. de Fermat (optical) and G.W. Leibniz, and that was developed in mathematics by L. Euler and J.-L. Lagrange, and continued by W.R. Hamilton.

---

32 According to the expression by Murray (1926), of which we have seen the branching law, optimizing the flow of fluid transported in a circulatory system (blood vessels, xylem of plants).

The importance of this notion of control, fundamental in physiology, is at the foundation, on the one hand, of *cybernetics* and, on the other hand, of the *optimal command of systems*. This, widely used in physics, begins to take place in biology for various applied problems. For example, in the higher plant, modeling of the transition from vegetation development to reproductive development that optimizes biomass from a given compartment (production of seeds, for example, rather than vegetative organs) or management of a given bio-industrial process, or even an aid to controlling the development of parasites in a culture. In this field, the essential notions that the biologist must use are based on the classic formalisms of L. Euler and Lagrange, renewed by Hamilton (1833). In fact, the development of these variational methods oriented towards an optimal command of systems by the set of control variables was permitted by the research carried out by the Russian school of thought in the 1950s, with Pontryagin's principle of maximums, specifying the necessary conditions for optimality[33].

3.6. Introduction of the formalism of dynamic systems in biology

Analysis of the development of multispecific populations, whose diversity of behavior and in particular whose temporal stability was already known, was considered from the point of view of mathematics in the period 1918–1939 and became an important research topic. During this era, seen as the "golden age of theoretical ecology", A.J. Lotka (1880–1949) and V. Volterra (1860–1940) described simultaneously, but independently from each other, the mathematical bases of the **dynamic of populations**. Let us cite their fundamental books:

– by Lotka: *Elements of Physical Biology* (1925), *Elements of Mathematical Biology* (1956), *Analytical Theory of Biological Populations* (1934);

– by Volterra: *Leçons sur la théorie mathématique de la lutte pour la vie* (1931, republished in 1990), *Principes de Biologie mathématique*, published in *Acta Biotheoretica* (1937).

The principle of these studies consists of postulating a system of equations for the speed of variation of the population numbers (or of biomass) of each species in association. The innovation is not simplified by using a differential formalism, where the latter was already used as a foundation of the first growth laws. In any case, the dynamic of populations was the first large field of biology that could then be understood from the point of view of formal principles of statistical mechanics. Effectively, in this field that is so rich in varied observations, it was possible to establish the existence of an invariant which was maintained throughout a

---

[33] In the case of a linear system, Pontryagin's principle establishes a necessary *and* sufficient condition for optimal control. See (Cherruault 1983, p. 80 *sq*).

conservation process (first integral, in the sense of the equivalent of the constancy of movement in mechanics).

But another new aspect needs to be pointed out, which relates to the conjugation of two components that are supposed to represent, on the one hand, the potential or "intrinsic" growth of each species (= as if it were alone) and, on the other hand, the interactions between species. Thus, by taking prey–predator models as examples, we are dealing with two types of interactions, intraspecific (competition) and interspecific (predation or parasitism), i.e. in the additive form for species i in association with species j:

$$\frac{dx_i}{dt} = f_i(x_i) + g_i(x_i, x_j) \qquad [3.1]$$

In other words, we now know how to treat a biological association as a sort of system in the integrated sense (not simply additive) of this term, i.e. the sense that takes into account the existence of interactions between constitutive elements.

First, the method consists of looking for the stationary point(s) defined by $dx_i/dt = 0$, then of determining their stability conditions. A great diversity of dynamic behavior can present itself (see Chapter 5). Finally, we need to verify whether the stability properties are modified or not following a slight variation in parameters of [3.1]. If they are conserved qualitatively, we talk about "structural stability".

This methodology was illustrated with many examples of various types of biological associations (prey–predator, species in competition, mutualism). Thus, this pioneering work led to an explanation of the decisive conditions, for example the exclusion of a species versus maintenance of a coexistence or the occurrence of an oscillatory behavior and its characteristics. Following this, many models continue to be perfected depending on the situations, as a function of the type of potential growth (exponential or logistical, for example) and of the type of interactions. Thus, for prey–predator systems, there are well-documented quantitative relationships that have been validated with various types of responses known as "functional" (consumption of prey per unit of predator) and "numerical" (specific speed of the predator).

The significance of this work easily surpasses this framework of theoretical ecology, because following this, the differential formalism occupied a prime position in biology. The methodology of dynamic systems (differential equations or equations with partial derivatives) thus became a basic tool for studying all kinds of processes.

We note that a discrete formalism (recurrence equations) presented itself a little while afterwards, in concurrence with the differential systems by Lotka and Volterra, with models known as "matrix models", developed subsequent to P. Leslie (1942). These became a tool of choice in the dynamics of populations organized into classes based on a given criteria: age (matrices by Leslie), size (matrices by Usher) or any other criteria of state.

It is important to highlight the difference between this approach by differential equations and models of statistical biometry that we have seen be developed from the first experimental designs in agronomy. We do indeed note that Fisherian probabilistic models are focused, in themselves and via analyses of variance, on the observed effect more than on the cause. According to their language, their objective is to "control" the variability of observations by considering the probability of repetitions. From an explanatory point of view, we can say that this is passive control: controlling is then calculating *a posteriori* rather than truly mastering. Moreover, we know that this term "control" can have various meanings (*control* also meaning control value or lack of treatment), in particular in the field of the "optimal control" of processes (see section 5.3.4.3), where the term "command" is often preferred (to clearly mean action).

It is more important to keep in mind that the approach by Lotka and Volterra has the characteristic of being based on hypotheses set down *a priori* and which define how a conjugation is set up quantitatively between potential growth and interactions that result from the association of several species. In other words, this is a prefiguration of a *systemic point of view*. Thus, some researchers talk about hypothetical or theoretical models to distinguish them from empirical models in statistical biometry. Let us explain this distinction by remarking that the hypotheses of these differential models in the dynamics of populations are not purely theoretical, in the sense that they can also rely on various empirical relationships. For example, the empirical formulation of different types of consumption of prey by predators depending on their abundance leads to an enrichment of the initial approach made by Lotka and Volterra, leading to a greater variety of dynamic behaviors.

3.7. Morphogenesis: the need for mathematics in the study of biological forms

Physiology and morphology are two extremely different fields in terms of their nature and their approach. Like many others, Bernard considered it important to carefully distinguish between them. In physiology, which has the advantage of being able to experiment, he opposed morphology, in which we can only scarcely observe because we are not masters of its determinism, which depends mainly on heredity, which escapes us. "We separate absolutely vital phenomenology, the objective of physiology, from organic morphology whose laws are studied by naturalists [...], but which escapes us experimentally and which is not within our reach"[34]. That is, these "laws of organic morphology" cannot be verified by experiments, and we are obliged to conclude that, in fact, "form is given in advance". Moreover, this last expression is revised, at least implicitly, by D'Arcy Thompson, a great observer of the living forms that sidelined heredity, and which we are going to examine.

In reality, while these remarks bear witness to an actual situation, of a sort of frequent divorce between these two branches of biology, they should no longer have a reason for being so. The external morphology is dependent on the internal functioning. Both of them need to be related at the same time to a complex genetic determinism and to the existence of various constraints, mechanical or other in nature (let us say epigenetic *lato sensu*). We will attempt to demonstrate through many examples the evidence for this position. For the moment, let us examine how mathematics intervenes in the study of biological forms by focusing as much as possible on the principle of non-separation of the form, of the structure and of the functioning. We invoke Buffon who, implicitly, related the external form that he was studying (in the same way as a sculptor would) to the existence of an "internal mold" that he presumed to be generating or modifying.

In the extremely vast field of biological morphogenesis, two different routes were drawn up, corresponding either to an overall study of forms (morphometry) or to an analysis of their dynamics (genesis and stability of morphological structures).

3.7.1. *General principles from D'Arcy Thompson*

Research into forms was historically influenced by the unique book by D'Arcy Thompson (1860–1948) *On Growth and Form* (two volumes, 1917, republished many times), who developed his main ideas using a great number of examples:

– "the form of organisms is directly determined by the action of physical forces";

– the principle of a mathematical transformation must allow different related forms to be connected. We show some well-known examples Figure 3.5, using changes in the coordinate system. Other examples are given on the variations of external morphology of the shell of crustaceans or even on those of the human skull.

---

[34] In Bernard, C. (1879). *Leçons sur les phénomènes de la vie communs aux animaux et aux végétaux*, vol. 2. Librairie J.-B. Baillère et fils, Paris.

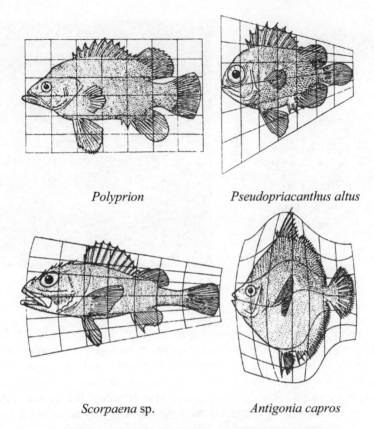

Polyprion *Pseudopriacanthus altus*

Scorpaena sp. *Antigonia capros*

Figure 3.5. *Examples of transformations of forms according to D'Arcy Thompson*

It is time to briefly recall, subsequent to D'Arcy Thompson, the mathematical description of certain forms of animal organisms (Figure 3.6), such as the shells of nautiluses or ammonites, based on the logarithmic spiral (or equiangular spiral):

$$r = ab^\theta; \quad a \text{ and } b > 0$$
$$= a\exp(m\theta); \quad \log(r) = \log(a) + m\theta$$

The length of the radius vector increases in a geometric progression when it sweeps through successive equal angles.

Shell of *Nautilus pompilius* After J.C. Chenu

Perisphinctus sp. ammonite

Figure 3.6. *Examples of biological forms with a logarithmic spiral*

It is useful to spend some time thinking about the way in which D'Arcy Thompson approached the role of mathematics in biology[35]. Its primary role, he said, arises from their fundamental propensity to produce and manipulate symbols, thus allowing a simple description to be condensed in natural language, i.e. by "economizing thought". He explicitly quotes (pp. 267–269) the famous words of Galileo concerning nature expressed in mathematical language, going back to Plato's idea of "God as a geometrist". In response to those who see in mathematics a series of excessively rigid definitions, he responds, with a certain degree of lyricism, that "it is precisely in its rigor that its infinite liberty lies", making reference to the case of conics whose definition expressed by their general equation (i.e. an invariant) elegantly encompasses various types of curves of very different "facies". These curves are thus found to be grouped together by the property of *homology*. Mathematical study of forms illustrates this power of mathematics as a tool that is both descriptive and analytical.

This important notion of homology led D'Arcy Thompson to the meaning of the concept of form that had therefore been mathematized, resulting in a two-fold comparison between forms, both static and dynamic. Thus, he confirms, we can move "towards the comprehension of forces" which has given rise to the morphology that is observed. While waiting to reach this ambitious objective of "comprehension of forces", D'Arcy Thompson explicitly asks the question: what is

---

35 We refer here to the quotes of Chapter 9 of the French translation of *Forme et croissance*. Le Seuil, Paris, 1994.

a change of form if not the result of "a movement of matter"? Here, we can see the implicit announcement of the notion of what we will later call a *"field of growth"* whose analysis will develop from the 1940s onwards with the use of vector analysis to correctly define the local growth activity in terms of intensity and direction.

Finally, another role of mathematics is mentioned, associated by D'Arcy Thompson with H. Poincaré, meaning "the very particular power of mathematics: combining and generalizing". It is this power that makes mathematics appropriate for the study of natural phenomena that are naturally composite in nature. Thus, "growth and form are entirely composite in nature", calling for the use of mathematical models that allow "large amounts of phenomena or more elementary actions" to be processed. Of course, today this statement needs to be revised with the advent of the notion of complexity which places the focus not only on the sum in the sense of the simple addition of components, but also on the importance of their interactions. D'Arcy Thompson adds to this (without developing it) another point that is not at all secondary, in the form of a piece of advice given to biologists (always faced with the variability of what they observe), which is to "learn from mathematicians to eliminate, to move away from and only keep the essential notion in order to reject what is particular, [...] sacrificing what is accessory in favor of what is typical" (p. 269).

We see that these pages by D'Arcy Thompson reveal in their style a new way of thinking that must modify the analysis of biological morphogenesis. The fundamental importance that he attributes, sometimes with lyricism, to the position of mathematics in biology does not at all prevent him from having a sense of limits. Thus, he notes that "there are, even within physical sciences, a certain number of problems that are outside the scope of mathematics today". In any case, many commentators do indeed underline the rather paradoxical destiny of D'Arcy Thompson's work. His book is in fact often cited as a pioneering book established on the basis of a broad culture. However, although it constitutes a source of inspiration for many problems of biological form, its direct applications are still considered to be very restricted. It is also useful to add the following remarks.

Indeed, D'Arcy Thompson's ideas about the search for a *coordinate transformation system* to formalize changes in form are not at all unknown. Better than that, they were revised and widely developed, particularly in the United States by F.L. Bookstein who dedicated a large amount of work in the decade 1980–1990[36] to morphometry, considered, according to his words, as an "empirical fusion of geometry with biology". It is an association of two sources of data: (i) the

---

36 Bookstein, F.L. (1978). The measurement of biological shape and shape change. *Lecture Notes Biomath.*, 24; Bookstein, F.L. (1996). Biometrics, biomathematics and the morphometric synthesis. *Bull. Math. Biol.*, 58(2), 313–365.

geometrical position, with respect to a given reference framework, of a set of reference points (*landmarks*); (ii) the biological homology or correspondence between these reference points whose bijection needs to be studied in order to analyze the transitions of form between related objects. Various applications have been made in anthropology and for medical purposes (morphology of the skull and the face) for which Bookstein carried out an important methodological renewal of D'Arcy Thompson's proposals. The analogy of changes in form induced by growth compared to the deformations of a solid due to mechanical constraints in a continuous environment was in fact postulated; a problem that is now treated using *tensor calculus*[37]. This methodology highlights the existence of principal axes of growth that determine the transition or movement of reference points. These directions are defined by eigen vectors of the growth tensor (symmetrical matrix made up of constraints in various directions). Independent of this general morphometrical work, we encounter this important notion of axes or principal planes of growth in the distribution of mitoses within a plant meristem, a distribution that is therefore directly determined by the properties of the growth tensor (Hejnowicz). To finish, let us add that all this applies either during development (ontogenesis) or between related species (phylogenesis). That is, at least for the latter point of view, this methodology does indeed correspond to the study of continuous, gradual variations, rather than abrupt variations in the form of leaps.

It is worthwhile giving a little more information about the significant remarks of D'Arcy Thompson when he mentioned the evident point about morphology, its reason for being, which consists in the end of *linking together the whole and its parts*.

> "Biologists and philosophers learn to recognize that a whole is not simply the sum of its parts [...]. It is not the case of a simple juxtaposition of parts, but of an organization of these parts, of a reciprocal arrangement of parts that are adapted to each other"[38].

While he, of course, sees in this something fundamental, which underlines the basis of the law of compensation and balanced growth by É. Geoffroy Saint-Hilaire, it is useful to note more generally that Thompson is one of the biologists who was able to prefigure (here restricting themselves to morphogenesis) the installation of an integrated or systemic biology that we will return to later on.

---

37 On the basis of the contribution of tensor calculus to the analysis of a morphogenesis, see Bookstein, F.L. (1984). A statistical method for biological shape comparisons. *J. Theor. Biol.*, 107(3), 475–520.
38 Thompson, D'A. (1992). *Forme et Croissance*. Le Seuil, Paris, 263.

REMARK.– At the margins of biological morphometry, as we have just illustrated, various studies are currently underway as part of what is known as the "*recognition of forms*". This rising theme currently comes more from bioinformatics than from explanatory mathematical modeling. Ranging from the level of cellular infrastructures to recognition of the form of organs, we can cite as an example certain software programs such as *MorphoLeaf*, which is designed for computer analysis of the variation of the form of leaf blades in plants. The morphology of the outline was studied on *Arabidopsis*[39], including the formation of teeth and sinuses that outline characteristic leaf lobes. This tool is dedicated particularly to the variations related to the growth and the position of the leaf on the carrier axis (heteroblasty). This research is therefore to be differentiated from a field analysis that is interested in the distribution of the local activity that induces a regionalization of the field and therefore of the form (see section 6.3).

3.7.2. Turing's reaction–diffusion systems (1952): morphogenesis, a "break of symmetry"

In biological morphology, another question is asked, which is to describe not a given form, but its appearance, considering that all morphogenesis is a generating process. From this point of view, very different from the previous notions of homology, a major innovation in the explanation of biological structurations was made by the mathematician and computer scientist A. Turing (1912–1954). It was explained in detail in his famous article from 1952: "The chemical basis of morphogenesis". The question that Turing considered was the following: how can a form, a structuration, appear in an embryo or a tissue that initially appears homogeneous, symmetrical? In other words, how can a morphogenesis be related to a fracture of symmetry?

While it is indeed necessary to refer to mechanisms that are chemical in nature by postulating the intervention of substances called morphogens, innovation mainly refers to the mathematical formalism used. Turing effectively postulates that the local differentiation of cellular structures can be expressed from *a system of differential equations* that specify (i) reactions between morphogens; and (ii) their unequal diffusion within an initially homogeneous tissue.

Due to the establishment of differential gradients between morphogens, the substrate becomes highly inhomogeneous (except for thermodynamic equilibrium). Native inhomogeneity can thus be broken following the occurrence of a random disturbance. Turing based his work on the analysis of the stability of his system

---

39 Biot, E. *et al.* (2016). Multiscale quantification of morphodynamics: MorphoLeaf software for 2D shape analysis. *Development*. 143(18), 3417–3428.

following this "break of symmetry". He shows that a precise distribution of morphogens can result from it just as it can correspond to the facies of certain biological structurations. The distribution of morphogens has the value of a "pre-pattern", a prior condition to a specific cellular differentiation. We are referring to **reaction–diffusion systems** and to **Turing's structures**, drawing a link between a continuous microscopic level (concentrations) and a discontinuous macroscopic level (discrete forms).

We can point out that the notion of morphogen was then just as hypothetical (and very often remains so) as the notion of gene at the origin of genetics. We will come back to the formalism of the differential equations used. Subsequently, Turing's principle of reaction–diffusion systems was widely used as a generally applied formalism that can describe carried morphogeneses, whether they are, for example, animal regeneration (with the famous case of the hydra), pigmentation motifs, or even the differentiation of new tissues or organs in plants and in vascular histogenesis or branching. Among the widely cited examples of spatio-temporal structurations, we can indicate the differentiation of pigmented zones on the epidermis of mammals or the shells of mollusks, as well as the Belousov–Zhabotinsky oscillatory reaction in chemistry. This methodology, now widespread, is thus appropriated to a certain type of emerging process.

3.8. The theory of automatons and cybernetics in biology

In competition with Turing's differential systems, the analysis of discrete systems took place in biology. The perception of numerous phenomena is indeed found in a population of physically distinct elements, such as the cells or modules of a spatially organized set. The study of cases of this kind can then be approached according to a principle known as "*cellularity*" (*lato sensu*), and not of concentration. In this context, biologists observe and measure the generation of new elements and their evolution, and not the existence of gradients or of singularities of a continuous variable.

3.8.1. *The theory of automatons*

At the origin of its link to biology, ideas were applied from the mathematician and computer scientist J. von Neumann (1903–1957), who sought to design *a machine that could self-reproduce*, the principle of which is known as "*cellular automatons*". These were popularized by the famous "game of life", proposed by the British mathematician J. Conway in 1970. He envisaged the format of a two-dimensional grid of boxes or cells, where each can take on two states, alive (1) or dead (0). The evolution of each cell is determined by simple deterministic rules as a function of its own stage and of the state of adjacent cells. J. Conway attempted several sets of rules, remaining particularly interested in the simulation of periodic structures. Here is an example with the following conditions:

– birth: from one dead cell with three live neighbors;

– survival: one living cell with three live neighbors;

– dead: one living cell that has less than two or more than three live neighbors dies (by "suffocation" or by "isolation").

We indicate here (Figure 3.7) three cases of development of a population of cells arranged as indicated, restricted to the future of the cell marked with a "?". A colored box means "living cell"; and an empty box means "dead cell".

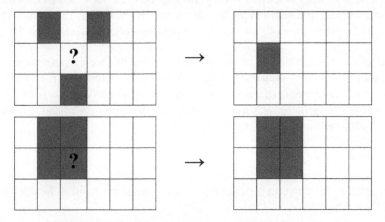

Figure 3.7a. *Simple example of Conway's game of life*

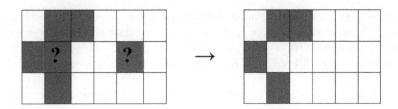

Figure 3.7b. *Simple example of Conway's game of life (second part)*

Stable structures filling a grid of this kind can be obtained with this type of cellular automaton. For example, with rules other than the previous ones, we can obtain the periodicity of the following motif:

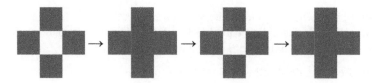

Figure 3.8. *Example of periodic structuring with Conway's game of life*

Over and above this particular device, chosen here due to its simplicity, the notion of automaton has experienced a development of its application in biology as a basis for the explanation or simulation of certain morphogeneses. Without necessarily being subject to a cellular substrate, like Conway's grid, the principle of automatons in biology consists of subordinating the behavior of a given element (a cell, for example) to a set of rules that determine its evolution. This principle is at the root of L-systems, well-known models in plant morphogenesis (section 3.8.3).

The question naturally arises of the comparison of this discrete approach to analysis by differential equations, where these are effective in emphasizing the properties of the dynamic of evolution of the system. A classic case is to compare the function of logistic growth in the continuous (function of Verhulst, Chapter 5), defined by the well-known differential equation:

$$dy/dt = a\,y\,(1-y/K)$$

which generally presents two stationary states: 0 (unstable) and K (stable), to the logistics known as discrete:

$$y(t+1) = a\left[1-y(t)\right]$$

This displays a varied dynamic depending on the value of the parameter a, evolving either towards a unique stable stationary state, or towards a regular oscillatory regime (with a variable number of cycles), or even towards the installation of a chaos whose detail is conditioned by the initial conditions.

This latter function is an entirely classic example of bifurcation, meaning an abrupt change of dynamic at each of the threshold values of a given parameter. Although the notion of bifurcation is general in scope, because it is observed in many multistationary differential systems (existence of attracting sets), its occurrence visualized by discrete logistics is a good illustration of the possibility and the importance of an evolving behavior. Thus, we are able to observe its importance, not for a simple elementary function, but for a discrete system of cellular automatons that is more or less widespread. Otherwise stated, we can reasonably expect to see certain discrete systems being able to exhibit, under certain conditions, a variation in dynamics that can carry it towards a chaotic behavior. We see the connotation with the principle of order carried by the property of self-organization.

Certain authors (such as T. Toffoli in 1994) thus considered that cellular automatons are more appropriate than the differential formalism to model an evolving physical system, since they can represent the underlying dynamic in a discrete form with the appearance of transitory situations without waiting for the completion of calculations of the overall dynamic (integration of a differential system). We will look at this delicate question in more detail with "complex" network systems (section 3.12.4). This subject, which has been widely studied since the 1990s, is of great interest in biology where the notions of order and stability can be deduced from the functioning of Boolean genetic networks (Kauffman in 1993).

3.8.2. *The contribution of cybernetics*

Since its beginnings, cybernetics has referred to biology, and its inventor N. Wiener defined it in 1947 as "the scientific study of control and communication in the animal and the machine". Although the term itself dates back to Antiquity with Plato and although there were precursors to retroaction mechanisms (like the adjustment system in a mill consisting of a crook string and a twist peg or the centrifugal governor by J. Watt, 1788), the advent of cybernetics constitutes a major stage in the study of the phenomena of regulation. Following the theory of communication by C. Shannon (1948), it is not a simple corollary, because it contributed a valuable addition to this due to the fact that it had been marked since its beginnings by the functional analogy that it set up between the activity of the nervous system and the control of a machine. In reality, a long time before this connotation was made by the neurologist W.R. Ashby (1952), physiology presented itself as a true prefiguration of cybernetics. It was effectively the discovery of hormones (by Bayliss and Starling in 1902) that clearly explained the principle of a causality defined by the "effector (hormone = excitation, beginning of motion) → remote receiver" couple. Later on, still without referring to cybernetics, the discovery of the effects of hormonal retroactions occurred.

The originality of cybernetics, according to the terms of Ashby, is to "not be concerned with objects but instead behaviors". It does not ask the question "what is this?" but "what does that produce?". It thus generalizes the notion of action: it is no longer just an action in the mechanical or physical sense, but of all types of signals (such as chemical, hormonal, or other actions) and more generally of all kinds of information.

On the contrary, the two following points characterizing cybernetics are added to the basic notions of automatons (state, inputs, outputs). Its founding idea effectively postulates that commands for all actions are internal (which obviously does not exclude the role of external agents). In addition, it adds a circular causality to direct unidirectional stimulus → response causality, as a condition of self-regulation: the effect acts on the source to control the operation (retroaction or *feedback*).

3.8.3. *The case of L-systems*

This remarkable type of cybernetics automatons was proposed and developed in 1968 by the Hungarian biologist A. Lindenmayer. The initial applications of this new formalism related to the growth and morphogenesis of single-series filamentous systems (1 row of cells in the same way as for certain species of algae or mushrooms), before being extended to branched systems. The *simulation of plant architectures*, including those of higher plants, then benefitted from this type of automaton that was developed in particular by P. Prusinkiewicz in the 1980s. Computer science tools dominate here, largely promoted in the book by Prusinkiewicz and Lindenmayer[40].

The formalism of L-systems arises from the research of the American linguist N. Chomsky in the 1950s into **formal grammars**. Applied to our subject, this term designates the set:

{vocabulary of cellular states, rules of production or syntax, initial state}

These grammars are described as generative, because we proceed in an iterative manner by re-writing the state of each cell at each stage of development (by stage we mean a step in the operation of the algorithm). For example, in the case of a single-series filament or of a tissue (cellular paving), each cell evolves according to its state and the various inputs received. Its development is thus determined, at least in part, by a set of cellular genealogies (notion of "ancestral memory", J. Lück, 1977). More generally, the principle of L-systems applies to a great variety of situations, in 2-D (cellular paving of a tissue) and in 3-D (embryogenesis, branched systems).

3.8.4. *Petri's networks*

This methodology from the mathematician C.A. Petri (1962) proposes a dynamic graphic tool designed originally for *qualitative modeling of discrete elements*.

---

[40] Prusinkiewicz P., Lindenmayer, A. (1990). *The Algorithmic Beauty of Plants.* Springer, New York.

Following this, extensions to these standard networks were developed for systems where certain variables can vary in a continuous manner (hybrid networks)[41]. This formalism began to be used in biology, in particular for the study of biochemical networks or networks of morphogenetic processes, for which it constitutes a useful stage in their study and that we are going to describe in broad terms. Without waiting for the more general question of networks (which will be envisaged later on), here we briefly examine what Petri's networks are, in comparison with L-systems, where each formalism has its specificities in the field of biological applications. A simple illustration of this is provided in Figure 3.9, limited to a single enzymatic reaction, glucose phosphorylation, the first stage of the phenomenon of glucolysis. This reaction is "discretized" according to a qualitative or logical point of view: the triggering of the reaction or not, without referring to the usual kinetic notions (molar concentrations, reaction speed). On this basis, we see the possibilities of extension to metabolic networks made up of variables of various natures, biochemical or genetic, and subject to various interactions.

Petri's networks are made up of two types of knots, denoted "places" and "transitions", linked by directed arcs (graphs). Each of the variables in the system in question is attributed to a given place that is described as "marked". Any change in state corresponds to the passage of a mark (or "counter") of one place to another, a passage that is subordinate to crossing a transition under a condition of validation. A transition is validated if various entry places are marked. Various cases can be presented depending on the existence of conflict between several transitions (when these have a shared entry point), or on the contrary, when there are structurally parallel transitions (no shared entry point).

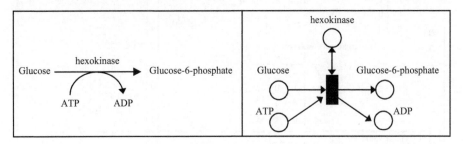

Figure 3.9. *Diagram of a simple Petri net, describing the reaction of glucose phosphorylation. The places are indicated by circles for each of the variables. The rectangle represents the transition, validated here by the presence of hexokinase that triggers the reaction (relation of influence noted by the double arrow)*

---

41 There is an active community that groups together researchers and users of Petri's networks (*Petri Nets World*). Website address: www.informatik.uni-hamburg.de/TGI/PetriNets.

The functioning of a Petri net consists of an evolution of the marking of different places via the crossing or execution of transitions. The formalism is based on (i) an initial vector M_0 (occupation or marking in different places); (ii) matrices of the arcs between places and transitions Pre = (P×T) and between transitions and places Post = (T×P). The state of a variable is specified, in a discrete manner, by the number of marks that occupy the place that is attributed to them. A transition is said to be "validated" (operational) if each of its entry places contains at least the number of marks required (meaning the weight of the arc involved), hence triggering of this transition of which the effect is to modify the marking of the entry and exit points, respectively by retreat and exit.

Figures 3.9 and 3.10 provide simple examples of standard Petri nets for discrete and non-temporized events. Let us point out, as an extension, the type known as "colored Petri net", in which we distinguish various types of marks that are specified or weighted by a given color.

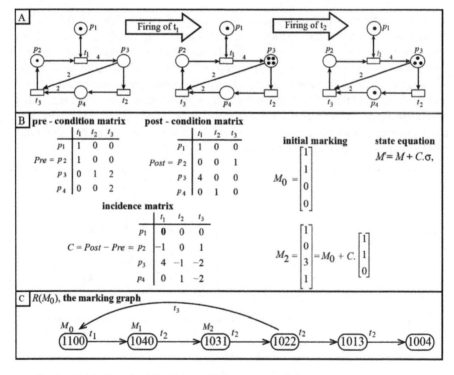

Figure 3.10. *Standard Petri net with four places (p) and three transitions (t)*

COMMENT ON FIGURE 3.10.– *The weight of arcs is specified, except if it is equal to 1. The particularity of place p1 is to correspond to an action of influence (unconsumable resource). Pre: preconditions to transitions (marking of entry places); post: productions (marking of exit places). Incidence matrix: assessment for each transition between productions and consumptions. The state equation indicates the markings on the net after the transitions have been crossed (according to the notion of the Parikh vector indicating the number of occurrences per transition; see Chaouiya, C. (2007). Petri net modeling of biological networks.* Briefings in Bioinformatics. *8(4), 210–219).*

Petri nets are a formalism well-adapted to the logical representation of a morphogenesis, meaning a set of discrete modifications (qualitative change of state or generation of a new element), stages that mark out the course of an ontogenesis. This is the case, for example, of a series of morphogenetic events that characterize the development in plants of a caulinary axis determined by the functioning of an apical meristem generating new modules, where these can be qualitatively different or not (depending on the passage to the reproductive state by sexualization of the apex). This method allows a global representation to be obtained of the structural relations between the development, simultaneous or delayed, of the various subsystems in the plant. For example, the phenomenology of the development of a cutting of *Erica* related to the processes of organogenesis and growth at various sites (neoformation and root and leaf growth). Let us again cite the modeling of meristematic functioning of a plant axis using a Petri net, taking into account the interactions between the genes involved (WUS and CLV genes) at different levels of integration (stem cell, meristem, metamer). Each functional cycle of this net leads to the generation of a module on the stem[42].

From a more general point of view, we can draw a parallel between the two discrete formalisms, namely Petri nets and L-systems. Both methodologies stem from the theory of automatons. This clearly shows us, for example, their application to the simulation of various models of plant architectures[43]. L-systems, characterized by re-writing the new state at each iteration and/or neoformation of all the elements in the system, are entirely suitable for demonstrating genealogies and related cellular groups. This being the case, they are equivalent to spatialized models, a character that we only encounter very indirectly in Petri nets. These, apparently simpler in their vocabulary, highlight the validity of transitions, hence a more explicit dynamic character that inspired them to study biological networks, independent of their own nature, whether molecular, cellular or of another type. In particular, it must be noted that Petri nets are based on relationships involving the consumption of

---

42 Barlow, P., Lück, J. (2007). *Rhythms in Plants.* Springer, New York, 219–242.
43 Prusinkiewicz, P., Remphrey, W.R. (2000). Characterization of architectural tree models using L-systems and Petri nets. In *The Tree 2000*, Labrecque, M. (ed.), 177–186.

resources/production, due to the existence of two types of nodes, places and transitions. Their function consists, essentially as it happens, of triggering a transition, i.e. an activation action, leaving to one side what would come from an action of inhibition. On the contrary, due to their qualitative nature, these networks are well-suited to the study of systems for which only a little data of kinetic nature is available (e.g. genetic networks). Since we are referring here to the field of standard Petri nets, let us say that the processes of −/+ regulation are currently better taken in account by another discrete formalism, that of models of kinetic logic by R. Thomas, to which the evolved Petri nets can moreover be related (Chapter 5).

These different methodologies intended for the graphical generation of biological forms are sometimes currently grouped together under the name of "informational morphogenesis" (rather than mathematical). A more general view of this, without being exhaustive, is presented by J.-L. Giavitto and A. Spicher[44], who justify this approach by taking up the principle: "calculating a form in order to understand it", a question of debate that is connected to the status of the representation and of simulation (see the end of Chapter 2).

3.9. Molecular biology

Among the new ideas that have been set out in the above sections and their contribution to the advent of systemic biology, which was an indirect result of them, it is necessary to focus a little on the subject of molecular biology to illustrate the interlinked nature of these different points of view. Due to its fundamentally reductionist nature, it was inevitable that this new biology would lead to a large number of reactions that contributed specifically to a change of paradigm, after a rather euphoric period marked with undeniable innovative results.

This discipline, thus named by W. Weaver in 1938, naturally arose from a coherent series of concepts and data: the notion of "gene" (theoretical entity implicitly postulated by Mendel in 1865), the discovery of chromosomes as the carriers of genes (Morgan, 1866–1945), then their location on DNA (deoxyribonucleic acid) and not on proteins (by Avery, MacLeod and McCarthy 1944), the double helix structure of DNA (Watson and Crick 1953) and lastly the nature and functioning of "genetic determinants" at a higher level of organization, that of operons or coordinated systems of several genes that intervene for a given function (Jacob and Monod 1961).

---

44 Bourgine, P., Lesne, A. (eds) (2006). *Morphogenèse. L'origine des formes*. Belin, Paris.

This disciplinary field initially developed due to the great interest that several physicists were showing for the problem of heredity. Concerning this momentum, which has been widely discussed elsewhere, we will limit ourselves to citing, among other important names, those of N. Bohr (1885–1962) and above all of M. Delbrück (1906–1981) and E. Schrödinger (1867–1961). Delbrück thus participated in setting up a "phage group" from which bacterial genetics originated. The importance of this group at the time was more to do with its original approach than with the results obtained in the replication of bacteriophages without the use of their biochemical information[45]. Concerning this historic contact between physicists and biologists, we can take note of Delbrück's comments in 1949 on the surprise experienced by physicists on coming into contact with living things, a surprise that he attributed to "the lack of absolute phenomena in biology", because "everything depends on the time and place [...] each living cell carries within it billions of years of experience from its ancestors"[46].

Regarding this critical era in the history of biology, we should remember that the contribution made by physics related to the ideas produced much more than the concrete results that arose from them, no doubt except in the case of the physicist and chemist L. Pauling (1901–1994), with his work on the explanation of sickle cell anemia (deformation of red blood cells) described as a "molecular disease". Thus, in 1954, G. Gamov proposed the idea of a genetic code (a little after the discovery of the structure of DNA in 1953) as a simple hypothesis establishing a necessary link between DNA and proteins, but without providing it with experimental validation or precise correspondence to the observation. Here, we wish to point out that this conceptual renewal that biology enjoyed can in part be related to the mathematical background of these various physicists who were first and foremost theoreticians. To this end, their objective was to look for a representation that was both unifying and universal, able to bring together various distinct phenomena. Their motivation was to see biology achieve what physics had already acquired by this time, the mid-20th Century. This was the exact thought of Bohr for whom it was necessary to "carry out an 'epistemological transfer', to find out how the new vision of the physical world modified the view of living things"[47].

Let us return for an instant to Schrödinger's point of view that he developed in his famous book *What Is Life?* (1944), to which a suitable subtitle could be "From physics to biology" since it is emblematic of this period in which many physicists demonstrated their interest in biology. Taking inspiration from statistical physics (the representation of a system with numerous microscopic variables using a small number of macroscopic values), Schrödinger aimed to describe heredity in terms of

---

45 Morange, M. (2003). *Histoire de la biologie moléculaire*. La Découverte, Paris.
46 Cited in Jacrot, B. *et al.* (2006). *Physique et biologie : une interdisciplinarité complexe.* EDP Sciences, Les Ulis, 24.
47 Morange, M. (2003). *Histoire de la biologie moléculaire*. La Découverte, Paris.

molecular structure and thermodynamic stability, a novel position in biology at the time. Among the ideas he proposed about genetic code, we will not retain the detail of his hypotheses, but instead the necessity of the existence of genetic code that he sets out as a condition for maintaining the order that characterizes each living organism. Let us summarize what he said about the notion of a gene, far in advance of the discovery of the structure of DNA: (i) "chromosomal fiber contains the entire future of the organism, expressed in a sort of miniature code"; (ii) genetic information is included in the configuration of covalent links at the heart of a series of given molecules. This is not a monotonous series of a single "motif" that is reproduced identically, but instead an ordered arrangement of several motifs (an "aperiodic crystal") that provides the possibility to code a very large number of possibilities. Concerning this biological question, Schrödinger clearly points out his thinking: "the existing order demonstrates the ability of self-maintenance and of the production of ordered events" (Chapter 7, p. 183), which refers, we should say, to a principle of autonomy. The principle of "order from disorder", which is often on the agenda at the moment, seems in his view and, on the contrary, to be completely foreign to life (Chapter 7, p. 188).

Another essential motivation that supported Schrödinger's thinking was the conviction that it was necessary to go beyond structural aspects of Delbrück's molecular model. He effectively postulates that the code must be "in exact correspondence with a very complex and very elaborate development plan and must contain at the same time the means to put it into practice" (Chapter 5, p. 153). Yet Delbrück's model "seems to contain no indication of how the hereditary substance accomplishes its function", hence its call to an association with "biochemistry guided by physiology and genetics" (Chapter 6, p. 165). He concluded that:

> "Living substances, whilst not eluding the 'laws of physics' as they have been known to us until now, probably depend 'on other laws of physics' that have been unknown until now, but which, once revealed, will constitute an integral part of this science in the same way as the first ones" (Chapter 6, p. 166).

The theoretical considerations made by Schrödinger are often cited as an epistemological historical reference of the connection between physics and biology. At this level, they have had a significant influence, as recognized by the physicist J.A. Crick (1918–2004), coauthor with the biologist J. Watson of the double helix molecular structure of DNA. On this well-known subject, we underline the major importance now attributed to considerations that are both *metric* and *topological* (number and connections of nucleic bases, their spatial arrangement) that Canguilhem summarizes in order to achieve a good understanding of what has now become knowledge of living things.

Following the discovery of the structure of DNA, there was some curiosity about a "central dogma", an expression by Crick who wanted to indicate the importance of this new paradigm. This designation, which remained in use for a long time, refers to the following unidirectional linear transition, which has a major explanatory role:

DNA segment (locus) → transcription onto the RNA messenger → synthesis of a protein (translation) → function

In the following section, we present a few major points regarding the interpretation of this fundamental representation of molecular biology in order to fully appreciate its scope in relation to our objective of the formalization of living things.

3.9.1. *On genetic information*

First, we clearly see that this representation questions the status of this genetic determinism in terms of *causality*[48]. In other words, what do we understand exactly by "genetic information" and what is its role? Of course, genes *lato sensu* can be seen as a direct cause, without retroaction of the effect produced (such as protein synthesis) on the cause. However, this cause is specific in that it is a form of instruction that is intended to maintain morphophysiological conformity with the characteristics of the species. We express this by saying: in giving an order to work, a gene in fact acts as a standard, signifying this particular characteristic that a gene is not like other types of causality.

This genetic determinism operates via the execution of a program, meaning an *algorithm* that connects a set of instructions or procedures in computer science. The introduction of this idea in biology (see Chapter 1) thus creates another question: by "genetic program", do we mean a relevant biological concept or do we use it as a simple metaphor? Is it an artificial comparison or, on the contrary, a useful analogy? We will only cite the suggestion made by Mayr concerning the advantage of putting the language of molecular biology in relation to the language of computer science. Here, we underline instead the more serious and more precise critiques that were presented by H. Atlan on the principle of simplifying the action of genes to acting only as triggers for instructions in the program. These critiques concerned the development of biology in particular[49].

---

[48] Concerning this paragraph and the following, see Maurel, M.-C., Miquel, P.-A. (2001). *Programme génétique : concept biologique ou métaphore ?*. Éditions Kimé, Paris, 40 *sq*.
[49] Atlan, H. (1999). *La Fin du "tout génétique" ? Vers de nouveaux paradigmes en biologie*. INRA editions, Versailles. See in particular the section "L'ADN : programme ou données ?", 32 *sq*.

We summarize by stating that, according to Atlan, genes are both work instructions in computer science *and* data that can be used elsewhere. On the one hand, their action is indeed located on a program. However, on the other hand, they *also* intervene as part of the machinery of the cellular metabolism. This new position relies on the fact that there is uniqueness, invariance of a genome within a given organism, whereas there is a great diversity of structures and functions of one cell or of one tissue to another. In addition, we know through genetic engineering that any cell has the capacity to read any DNA (even from another species) and to execute its instructions.

The essential argument is that the implementation of a program in computer science implies the requirement for a very close link with an interpreter who can read it and execute it[50]. It is not sufficient to talk about a program of coded instructions; it is also necessary to have a device that is capable of deciphering it in order to implement it. While DNA is strictly a program, its interpreter must be located within its cellular machinery. While, on the contrary, DNA plays the role of data, the use of it implies that the program of instructions is located in the framework of a system of biochemical reactions (metabolisms and transports), a system that Atlan considers to be a Boolean network of automatons. In this case, the program presents itself as distributed, in the same way as a parallel computer. Of course, these two points of view are truly complementary.

However, regarding this concept of the operational sharing of instructions, we are required to consider some questions. Effectively, from the computer science point of view (logical) in which we place ourselves here, the status of program that is attributed to cellular machinery leads to consideration of the latter as a Boolean system (cellular automatons). Yet can we state that a cell behaves exactly like a given logic network? The question does indeed arise, because biochemical networks evolve over the course of ontogenesis (time-dependence), leading to a connectivity between elements in the system that is not permanently fixed[51].

With this revised viewpoint of molecular biology that we have just described, we see that processing of genetic information, as the determinism of the functioning of an organism, begins to resemble a computer science approach rather than referring to concepts that are strictly mathematical such as those associated with system dynamics.

---

50 In a highly schematic explanation: in contrast to the compiler, the interpreter allows the program to be executed by reading it instruction by instruction.
51 A brief comment is given in Miquel, P.-A. (2007). *Qu'est-ce que la vie ?*. Vrin, Paris, 126.

Another current aspect of molecular biology needs to be mentioned, resulting from the need to process a large mass of varied data, pertaining both to the genome itself and to its productions, transcriptome and proteome. Due to this diversity of nature and to the interactions at play, analysis of biochemical nets of this type now benefits from the methodology of networks, such as Petri nets, a formalism that was developed in 1962 (see section 3.8.4).

3.9.2. *The linguistic model in biology*

Let us recall that Schrödinger (1944) believed that chromosomes contain the entire future of the function and development of living organisms in the form of what he called a "script code". In this way, he anticipated the notions of genetic information (without using the term information which was first used by Shannon in 1948) and of genetic programming[52]. Heredity was becoming a question of information formatted in codes and messages, whose elementary operational unit consisted of a combination of four chemical radicals (the bases of nucleotides) arranged in the form of triplets (codons) that are valid as code for protein synthesis.

Here, we have seen an *analogy between heredity and linguistics*. Thus, the linguist R. Jakobson sought to bring together genetic code and verbal code, both based on a combination of elements (respectively chemical radicals and phonemes), from which their meaning is drawn. The term "linguistic model" in biology corresponds to this formal analogy on which Jacob made an interesting comment in 1974[53]. First, it is necessary to note that "formation of the complex through combination of the simple, hierarchized levels of construction through successive integrations of units of a lower rank do not only make up linguistic and genetic systems". The common factor between the latter is the *linearity of the structures* that they create. The distinction between the verbal and the genetic is found in the fact that genetic material has two distinct roles: the functioning of the organism and hereditary transmission. While the linearity of the genetic message allows (via the translation) various functions to be controlled, it cannot lead to a reproduction of structures in 3D. On this point, Jacob recalls the idea of an *internal mold*, proposed by Buffon as a necessary condition for reproduction in 3D (a qualifier that was badly understood in its time, which intended to point out the difference as opposed to a sculptor's mold whereby the latter produces the imprint of the surface but not the internal structure). Using Buffon's terms: "Nature is capable of making molds from which it produces, not only the external figure, but also the internal form"[54].

---

52 See Pichot, A. (1999). *Histoire de la notion de gène*, Chapter VIII. Flammarion, Paris.
53 Jacob, F. (1974). Le modèle linguistique en biologie. *Critique*, 322, 197–205.
54 Cited by Jacob, F. (1974). Le modèle linguistique en biologie. *Critique*, 322, 93.

The advantage of the linguistic model as observed by Jacob thus lies in the fact that it applies both to the structure and the functions of genetic material. In any case, it played a deciding role in the explanation of mutations (assimilated to copy errors input by the copyist or the printer of the verbal message) as well as the punctuation, which ensures a significant discontinuity in a continuous nucleic chain.

3.10. Information and communication, important notions in biology

The term **information**, an intuitive notion of everyday life, corresponds to a basic concept in the study of communication between an emitter (or source) and a receiver, which are all terms that biology has seen in experimental reality. For example, genetic information (DNA → RNA → protein), conduction of a nervous signal and transmission of a hormone from its site of synthesis (endocrine gland) to its site of action (target). The notion of a source-sink couple, so familiar in physiology in various processes, arises from this framework.

From the point of view of terminology, we should note (Gayon 2018) that while the various usual terms that describe a communication process (message, translation, transcription, editing) are indeed like linguistic metaphors, this is not the case for the now ubiquitous use of the particular term "information" in general. In contrast to genetic information, which can be looked at in parallel with linguistics (see section 3.9.2), it is necessary to state a concept of a much more general scope that applies to a great range of phenomena in which this type of analogy does not at all occur. For example, a reference to information about fixing a hormone to its receptor needs to be considered as a transmission that does not operate according to a specific coding of a genetic determination type in which there is correspondence term by term between nucleotides and amino acids. It follows that currently, the concept of information is of very wide-reaching importance, in the same way as matter or energy[55].

By information, in its general sense, we understand a theoretical concept that arises from the notion of probability. As an attribute, its quantification is based on basic axioms of the calculation of probabilities, which convey a strong character of abstraction to it. That is, an event A of probability $P(A)$. Achievement of it constitutes a piece of information that is all the more important since its probability of occurrence is low (the achievement of a certain event brings no information!). This is expressed as a function known as the "quantity of information" in the form:

$$H(A) = f\left(\frac{1}{P(A)}\right)$$

---

55 See (Ricard 1999, 2001).

under the conditions:

- for a certain event: $\lim_{P \to 1} f(P) = 0$;

- for two independent events: $f(P_1 + P_2) = f(P_1) + f(P_2)$.

Hence, the choice of a logarithmic function for *f*.

This notion was developed by the mathematician and engineer Shannon in his theory of communication (1948), whose objective was to study the *quantity of information* contained within a message conveyed on a given communication route. Continuing from the previous subject, the principle consists of associating the emitter with a discrete random variable X in n states i of probability P_i, which correspond to the emission of a message containing n symbols. The quantity of information contained in this message is given by the *Shannon function*, known as "Shannon entropy":

$$H(x) = -\sum_{i=1}^{n} P_i \log_2 P_i \qquad [3.2]$$

This is the mathematical expectation of the variable $\log_2 P$. Using the *bit* (*binary unit*) as a unit (two possible values: 0 or 1), a logarithm with base 2 is chosen.

In the event of equality of probabilities $P_i = p = 1/n$, we have:

$$H(x) = -\log_2 p = \log_2 n$$

The quantity of information H is a measure of uncertainty. It is the number of questions with a binary response (yes/no) to ask at the source in order to reduce the uncertainty. We see its formal equivalence to the nearest sign in the entropy of thermodynamics, which leads to the name "negentropy" (or negative entropy) from Brillouin.

This definition is extended using quantities of *conditional information*: $H(B|A) = -\log_2 P(B|A)$ (transmission of B, where A has already been emitted) and of *mutual information*: $H(A,B) = H(A) - H(A|B)$ (joint transmission of A and of B).

Let us now consider the data set in a communication (letters of an alphabet, for example) which correspond to the emitter and the receptor and which are designated respectively X and Y.

That is, the value H(X, Y), known as joint entropy of the communication channel, which measures the uncertainty of the couples (x_i, y_i). H(X) and H(Y) represent the respective "self-information" X and Y, calculated as the average uncertainties of these two sets of data (alphabets), i.e. by averaging the expressions – $\log_2 p(x)$ and $[- \log_2 p(y)]$. During the communication, noise can be introduced, in such a way that what is effectively transferred corresponds to a mutual information denoted I(X, Y). Knowing that the mutual information of two random variables measures their dependence in terms of probability (not in the sense of a causality, but simply of a statistical correlation, such as the classic Bravais–Pearson coefficient r of a linear correlation), we have:

$$I(X,Y) = H(X) - H(X|Y) = H(Y) - H(Y|X) = H(X) + H(Y) - H(X,Y)$$

The sum of the self-information is in general greater than or equal to the information that transits:

$$H(X,Y) \le H(X) + H(Y)$$

(property known as "*subadditivity*")[56].

This purely formal notion of quantity of information constitutes a theoretical basis for studying biological systems that exhibit self-organization, and that we are going to discuss. From a methodological point of view[57], we note that the Boolean approach of communication circuits was introduced by the Shannon theory. Biology benefitted from this in the logical formalization of regulation phenomena (e.g. networks of genetic regulation, see Chapter 5).

3.11. The property of self-organization in biology

This theme of very general scope is inextricably linked to the notions of information and communication that we have just summarized as the basis of all organization of a set of connected elements. Initially attributed to physical systems, these notions rapidly became increasingly useful for biology. Let us give a brief presentation of them.

The term "self-organization" designates the property of a living system to manage by *itself* to reach a certain structural and functional organization, to make this grow and to maintain it throughout its ontogenesis. It is an intrinsic property of the system and is fundamentally different to the action of external forces alone

---

[56] Concerning these basic notions, see, for example, (Ricard 2006, Chapter 7).
[57] See the detailed review by Ricard (2006).

(which naturally can contribute to it). This is an important aspect of the general property of the autonomy of living things.

The very act of self-organization is in opposition to the trend of increasing thermodynamic entropy, i.e. to the evolution of the system towards equilibrium (homogeneity) or disorder. Let us say that this maintains a characteristic order despite or thanks to the multiple causes of variation.

With the Austrian physicists H. Quastler (1908–1963) and H. von Foerster (1911–2002), the importance of this notion in biology took shape. It was also at the origin of cybernetics, in reference to Ashby. With Turing and von Neumann, the era was a prosperous time for new concepts and provided renewed thinking in biology.

3.11.1. *Structural self-organization*

The use of Shannon's quantity of information first gives rise to a static view of a structural self-organization (Quastler, *The Emergence of Biological Organization*). Quastler and S. Dancoff (1953) applied Shannon's theory to genetics to study the occurrence of errors in the transmission of genetic information.

The opinion of von Foerster (*Self-Organizing Systems*, 1960), different to that of Quastler, advocated an operational self-organization in which time intervenes on the variation of the interdependence of the elements of a system. In addition, he conceded that the environment can act on these interactions, focusing particularly on the importance of random fluctuations. This is the principle of the "creation of order through noise", for which mathematical theory was specified by the physicist H.P. Yockey (1958, 1974) who applied it to biology, in particular to the problem of the origin of life. This theory was finalized by the biophysicist and doctor Atlan (*L'Organisation biologique et la théorie de l'information*, 1972, 1992)[58]. According to von Foerster, self-organization has been repeated by numerous authors, such as I. Prigogine and collaborators on the thermodynamics of irreversible processes.

The idea that this principle of self-organization causes joint interaction between internal interactions and the action of the environment is fundamentally different from the idea of Schrödinger (1944), which has already been described, who believed that in living systems, there is a "formation of order from a pre-established order".

---

58 This theme was the subject of a lively controversy with Thom. See, for example, his comment: Thom, R. (1980). Halte au hasard, silence au bruit. *Le Débat*, 3, 119–132.

3.11.2. Self-reproductive hypercycle

An important theoretical contribution was added to these classic considerations with the concept of the self-reproductive hypercycle, imagined by the German biophysicist M. Eigen (1971). A hypercycle is a cyclic system with an enzymatic function that connects a series of self-replicating units (autocatalysis) in relation to external entries. Its operation can be formulated by a dynamic system of differential equations that couple the various elements at stake.

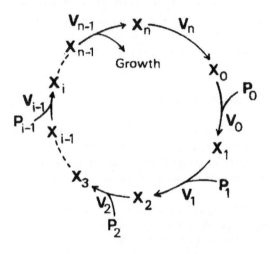

Figure 3.11. Model of a hypercycle that represents the ionic regulation of the multienzyme system of plant cell walls (Ricard and Noat 1986). P_i: carbohydrates incorporated into the wall under the action of enzymes V_i; X_i: intermediate compounds. The transition $X_n \to X_0$ is associated with the increase in the density of charge, after it has been reduced by growth triggering during the stage: $X_{n-1} \to X_n$

Its application to RNA has been proposed as an explanation to the self-organization of prebiotic systems[59]. Two key points are envisaged in particular in this new context: the emergence of a cellular structure with membrane, then the advancement to multicellular organisms[60]. Let us illustrate this concept with a model of regulation of the growth of plant cells (Figure 3.11). The contribution made by the application of the principle of hypercycle to this is being able to put into interaction the various elements of a complex set that is characterized by the temporal coordination of multienzyme activity, of the progressive incorporation of the structural elements that constitute the wall and of the variation of the ionic

---

[59] Eigen, M., Schuster, P. (1977). *The Hypercycle. A Principle of Natural Self-Organization.* Springer, New York.
[60] For an overview, see Ricard (1999, pp. 333–352).

charge of the latter. Growth is subordinate to a given stage of this cycle, which is determined by the state of the parietal electrostatic potential.

The principle of this structured arrangement of various elements in a circuit is used in addition to many other phenomena. For example, the system *Evolon* is associated with it, coupling an ordered series of various components that are involved in the same growth phenomenon[61]. This system provides an interesting phenomenological formalization of multi-agent situations in which the autocatalytic replication of elements and the set of their interactions are coupled together. This provides a basis for the interpretation of various biological processes that are usually understood via a global univariate mathematical function. For example, certain growth functions $y(t) = f(t)$ are explained as a structured set of interdependent elementary compounds $\sum_{j,k} x_j \, x_k$ [62].

3.12. Systemic biology

In order to develop what is represented by this new theme, we need to remember that its genesis results from the eruption, which occurred more or less jointly in biology, of the theory of automatons and notions of cybernetics, as we have pointed out above. Let us add some essential considerations that explain their connections despite the differences in motivation.

3.12.1. *On the notion of system*

If we refer to the property of autonomy of living things, we do indeed see that a biological system has the characteristics of Prigogine's dissipative structures, meaning structures that are maintained far from the state of equilibrium (therefore of a homogeneous state without local differentiations) by a flow of matter and energy, a flow which is controlled, at least in part, by the system itself. The term "equilibrium", in the thermodynamic sense, means the homogeneity of the various state variables and consequently a lack of structuring. Thus, we reach this new name for "biology of systems" or "systemic biology", whose current success testifies to a salutary reaction to the necessarily reductionist nature of biology that is attracted to the supremacy, if not the almost-exclusivity, of the molecular. Whether molecules are constituents of a living thing (since they are themselves made up of atoms and these again of elementary particles), it obviously does not mean that we can reduce it to an elementary level of organization. In addition, the intertwined nature of

---

[61] Peschel, M., Mende, W. (1986). *The Predator–Prey Model: Do We Live in a Volterra World?*. Springer, New York.
[62] See, for example, Buis, R. (2016). *Biomathématiques de la croissance*. EDP Sciences, Les Ulis, 459 *sq*.

several layers of organization evidently reinforces the weight of this term of system whose meaning needs to be specified a little.

The development of this trend of thought needs to be related to the work of the Austrian biologist L. von Bertalanffy, considered to be the founder of systemics. Returning to various previous research publications, the scope of his pioneering book (*General System Theory*, 1968) overtook biology alone, presenting itself as a universal paradigm (which inspired social sciences). Let us note that the contribution made by von Bertalanffy to biology, by insisting on the idea that biological systems are open systems (implying that they take input and output flows into account), relates in particular to the modeling of allometric relations and biological growth. Concerning this point, with respect to many classic formulations, its originality is to connect the laws of biological growth to the existence of various metabolic types.

3.12.2. *Essay in relational biology*

Within the sphere of influence of ideas that included the theory of automatons and cybernetics, on the one hand, and those of von Bertalanffy, on the other hand, concerning the very principle of what a system in biology can be, a movement was drawn up in the 1950s–1960s under the motivation of theoretical biologists N. Rashevsky (1899–1972) and R. Rosen (1934–1998). Their numerous works brought about a new current that they designated "relational biology".

This description needs to be justified, because it obviously does not surprise any biologist who is trained in the physiological thinking of Bernard, for whom everything is a relation in a living organism.

The name effectively needs to be understood in the mathematical sense of the *theory of categories*, which developed from the proposals made in the 1940s by the mathematicians S. Eilenberg and S. MacLane. Let us summarize the basic ideas:

– by "category", we understand a specific collection of objects {A, B,...} whose relationships we seek to formalize by a transformation or projection (*mapping*) in a given direction: A → B ;

– a function *f* assigns to each ordered pair of objects (A, B) a projection denoted H(A, B): $(A,B) \xrightarrow{f} H(A,B)$.

The sets of projections H(A, B) constitute an essential characteristic of the structure of the category in question.

Having correctly specified the objects in question and the transformations that they will undergo[63], these basic principles are completed by the classic axioms of composition and associativity, meaning: (i) a composition function that assigns a transformation denoted gf (composed of g and f) in H(A, C), to pairs (f, g) of transformations such as $f \in$ H(A, B) and $g \in$ H(B, C); and (ii) a property of associativity: if $f \in$ H(A, B), $g \in$ H(B, C), $h \in$ H(C, D), then $h(gf) = (hg)f$.

Generally, this theory aims to look at the existing relationships between various sets or different mathematical structures. The basic ideas that characterize this approach of establishing relations involve, on the one hand, the general notion of morphism (in the sense of the relations between two sets) and, on the other hand, what is known in algebraic topology as "homotopy" (pathway linking two structures). We will focus on the continuous transformation (in a given application) of an object or of an algebraic structure into another. For example, the transformation of a circle into an ellipse or the relationship between two vector spaces. We see the correspondence of this (in its principle) with everything that biology studies as part of interconnected metabolic and genetic networks according to a representation on directed graphs that are simply graphics of transformations. In other words, this theory focuses on the correlations between observables that have the nature of a morphism and that we seek to integrate into a diagram of categories.

In the simple transformation diagram between A and B that is determined by an application or projection f, the operation needs to be described in the following terms of causality:

$$f$$
$$\downarrow$$
$$A \dashrightarrow B$$

A is the material cause and f is the effective or driving cause (to repeat Aristotle's terms) of this transformation. This, for example, can correspond to a given stage in metabolism. Concerning this basic principle, Rosen developed the case of systems denoted *metabolism-repair* (M, R) which describe, at a cellular level, the relationships between two compartments: (i) cytoplasm (site of metabolic reactions); and (ii) nucleus (providing renewal, continuity of the metabolism). By "repair", he meant the reconstitution of components that have been destroyed or inhibited over the course of the metabolism. Since each reactional sequence has a given duration (*lifetime*), we can add a repair component Ri to each metabolic

---

[63] See Rosen, R. (1958). A relational theory of biological systems. *J. Math. Biophysics*, 20, 245–260 and 317–341; Varenne, F. (2013). *Rev. Hist. Sci*, 66, 167–197.

component Mi, where the components are connected to each other by a set of inputs-outputs. Through a representation of Rosen's idea, we will select the characteristic of these entities for our explanation, meaning: enzymatic for M, RNA for R and DNA for the projection that defines this transformation system.

Let us add an interesting theoretical point regarding what we call a "natural transformation" (or "equivalence")[64]. This name is related to a property of "functor" or application transforming a category into another. By "natural transformation", we mean the transformation of one functor into another, which respects the internal structure as it is defined by the law of composition of morphisms. Thus, in the very theoretical context of Rosen's ideas, we can imagine that a qualitative change of dynamic (bifurcation) has the characteristic of a natural transformation that would be determined by a given functor.

In his summarizing work in 1973, completed by his last essay, "Life itself" in 1991, Rosen highlights the analogy between his own representation in terms of algebraic topology and that of the interactive networks that relate to genetic regulations. He insisted in particular on the close underlying relationship that he saw between the graphical representation of the topological relations imagined by Rashevsky and himself, and the recognized relationship to the function of genetic determinants such as the lactose operon (set of genes that play a role in the use of this ose in cellular energy), which was an archetypical genetic network. During this search for theorization, another pioneering work was proposed by Rosen (1958) as the illustration of its method, that of the founding model of automatons by McCulloch–Pitts, which describes the logical properties of nerve conduction by formal neurons. It attributes to a "version" in the terms of its theorization of categories. Let us remark that this example is suitable for this, since the notion that is common to the two approaches is that of black boxes that are simply understood to be the flow of inputs and outputs, leaving to one side what is inside (which is related to what is called "the underlying mechanisms"). In this attempt to see the principle of this relational biology applied to phenomena that are physically highly diverse, Rosen obviously makes widespread use of the analogy, considering, for example, that in the formal unit of the operon, the same properties are found as those of excitable nervous elements. Unfortunately, these relations are scarcely explained in the same level of detail that a comparison between the two kinds (topological transformation and the dynamic of processes studied) could produce.

---

64 Concerning this theoretical question that has still not been applied much in biology, see Varenne, F. (2013). Théorie mathématique des catégories en biologie et notion d'équivalence naturelle chez Robert Rosen. *Rev. Hist. Sci.*, 66(1), 167–197.

Relying on the theory of categories conveys a highly abstract nature to this relational biology, which up to the present has not limited itself to its applications, an analogue observation that can moreover take place for other theories, such as the theory of catastrophes by Thom. Accepting the difficulty, Rosen sees in his undertaking the very principle of a true theorization in biology, which presents the dilemma: "is there a unified mathematical biology?"[65].

In other words, this relational biology would like to be more of a theory than a model. Maintaining that it is a guide to modeling, it is thought to have a much more general scope than the contribution from a specific model[66]. Nevertheless, more simply, the principles of this relational biology, by the topological nature that it presents to any biological transformation, constitute an interesting and original stage in mathematical biology connections, an approach that it is necessary to cite here in principle, given the paths taken today in light of a formalized representation of biological processes. We add that this work by Rosen was revised by D.C. Mikulecky (2007) as a theoretical basis for complexity in biology.

3.12.3. *Emergence and complexity*

In the face of inevitable limits to reductionism while maintaining a syncretic holism that is difficult to master and which can lack precision or coherence, there was a progressive increase in taking into account this property that is denoted **complexity**[67], an idea that is currently taking root but is not yet well-formalized in many scientific fields. To distinguish between complexity and complication[68], a *complex system* is defined as a set of elements in interaction whose behavior cannot be predictable with the individual properties of its elements alone. Let us say that the general definition of a system of this kind is completed in this way:

DEFINITION.– The most essential characteristic of a complex system lies in a property of non-additivity of the parts in the function of the whole, in such a way that the whole is more than the sum of its parts.

---

65 According to the title of the last instalment of the master work of which he is the editor: *Foundations of Mathematical Biology*, 3 volumes, 1973, Academic Press.
66 This question has long been debated by Varenne (Varenne 2010).
67 "Complex", an adjective, from the latin *complexus*, interwoven; from *complectere*, to embrace.
68 Let us specify that in mathematical/computer science terms, the algorithmic complexity or Kolmogorov complexity is a function that quantifies the size of the smallest necessary algorithm that can generate a series of characters.

We describe it as an *"emerging system"* in order to underline its property of exhibiting new properties (known as collective properties), which are different from those of its parts considered separately. It is not the number of elements that enter into the game, but the fact that a deciding role is attributed to the interactions and not the elements taken in isolation, which implies a change of point of view[69].

Here, it is necessary to correctly understand the difference between emergence and potentiality. This term "potentiality" is given the meaning of a powerful act by Aristotle, as we have seen. It is also favored by molecular biologists, such as J. Monod, whose point of view we have already seen and according to which the properties of all biological systems are already potentially present in the structure and the functions of their elements, and more precisely of their macromolecules, genome and proteome. Hence, the idea quite naturally occurs, associated with the principle of an organizing program, of knowing in detail the parts in order to know the system as a whole. Everything arises from this principle of executive code + program. In summary, it would mean going from a microscopic level of organization to a macroscopic level by simple additivity, combined with a program that exerts the role of an organizer. Thus, the genotype → phenotype transition is seen by Monod (1970) as an epigenetic construction which, except for all ideas of emergence, is equivalent to a simple "revelation" of what is already potentially present.

A first example, borrowed from biochemistry, is the well-known process of glycolysis. Effectively, we know that this essential metabolic route for the production of energy at a cellular level is capable of exhibiting a periodic behavior, whereas none of the enzymatic reactions at play present a property of this kind in isolation. The periodicity observed in the system has the nature of an emerging property. An entirely different example, borrowed from plant biology, can be noted here to illustrate the generality of this question. This is the description of growth of a stem that we could naively deduce, in the same way as a tautology, from the sum of its different constitutive internodes. Both experimentally and in reference to a mathematical model of growth, the sum of "local" kinetics cannot restitute all the properties of elongation of the global stem, in particular the existence of oscillations of the instantaneous speed of growth that each of the internodes does not at all exhibit. The whole is oscillatory, whereas its parts are not. Admitting the additivity of the local functions comes down to neglecting the non-synchronism of local growth. Effectively, each internode of rank $j + 1$ is only generated (let us say phenomenologically) when that of rank j acquires a certain age, meaning it reaches a certain stage of development. Although this consequence of a sequential organogenetic process is very different from the set of biochemical interactions that

---

[69] A presentation of the notion of complexity that emphasizes the contribution of the work of Rosen that we have just seen is given by (Mikulecky 1996).

occur in a chain or in a metabolic cycle, we can say that the spatio-temporal connection between the various internodes of the same axis has the same "explanatory" value here.

The origin of the term "emergence" dates back to the English philosopher and economist J. Stuart Mill (1843), for whom the appearance of an unexpected novelty, one that is known as emerging, meant an opposition to the "law of the composition of causes", because this is supposed to allow prediction based on known elementary causes. Stuart Mill forged the term of "laws" or of "heteropathic effects" to specify situations of this kind in which the observed effect does not correspond to the sum of the effects of elementary causes. The notion of emergence and the reflection of Stuart Mill were revised a little later by the English philosopher G.H. Lewes (1874) who, by relying on physical sciences[70], highlighted the difference between emerging facts (that cannot be predicted) and resulting facts (predictable from what precedes it). For his part, von Bertalanffy (1945) refers explicitly to emergence as a characteristic of complex systems.

This notion of emergence was rapidly accepted "in practice" by numerous English-speaking biologists. In particular, we cite the biochemist J. Needham or the embryologist and geneticist C.H. Waddington, both interested in a theoretical biology of morphogenesis[71]. For his part, the epistemologist C.D. Broad (1925) makes use of this to resolve the debate on vitalism. On the contrary, it appears that in France, the term "emergence" has not really entered the usual vocabulary of biology, whereas, meaning the contribution of an innovation, it is "something" essential in all biological processes of which we attempt to understand the dynamic, in particular via its singularities[72]. We can but give a reminder of how much the question is completely fundamental, "unavoidable", in these two particular domains of the theory of evolution and embryogenesis. In their respective contexts, it goes without saying that the demonstration of any innovation has a fundamental property value, whether it is a case of abrupt variations (not gradual) within a phylum or of the generation of a new structure at certain stages of an organism's development. Here, it is sufficient for us to cite the reference examples of the development of

---

[70] As a reminder, we can cite the famous argument of Lewes, who considered that the properties of the water molecule, different from those of its components H and O, must result from their interaction, therefore with the nature of an emergence.

[71] Waddington instigated meetings that brought together researchers from different fields, biologists and mathematicians (B. Goodwin, S. Kauffman, J. Maynard Smith, R. Thom, L. Volpert participated in this) in the 1960s. See the book that resulted from this, Waddington, C.H. (dir.) (1968). *Towards a Theoretical Biology*. Edinburgh University Press, Edinburgh.

[72] As a reminder, let us remember the point of view of Bernard; he pointed out the opposition between "life is creation", and the experimental demonstration by L. Pasteur of the impossibility of spontaneous generation of any kind.

the eye over the course of evolution or the formation of polarized axes that are well-defined at certain stages of embryogenesis, a major factor in any organization plan. Let us recall, for example, that in causal embryology, an important acquisition was the notion of an organizational center, a zone that needs to be defined in spatio-temporal terms for each of the differentiated outlines. Embryogenesis could thus be seen as a series of distinct morphogenetic inductions. Inclusion of a principle of self-organization in all ontogenesis, animal or plant, was later equated to a change of scale.

For any observation scale, emergence is presented as a discontinuity, in contrast with the continuum that is the underlying medium for it, let us say the form versus the basis or substrate. We see the connection with the mathematical notion of singularity, which refers back to the general continuous-discontinuous dilemma.

It is also a surprise to observe how little attention is attributed to this concept of emergence, or even the lack of it, in a range of works outlining thoughts on the general theme of "explaining life". Let us add that, even in epistemology, the importance of this notion was recognized in France late and with a certain degree of reticence. The classic reference dictionary by J.J.L. de Lalande demonstrates this (*Vocabulaire technique et critique de la philosophie*), in creating late on (in the 1940s) an entry for the term "emergence", and in addition only attributing a descriptive value to it, while refusing to see it in the sense of an explanatory hypothesis or a promise of intelligibility. This surprising characteristic is well-described by A. Fagot-Largeault[73], who discusses at length the meaning and the means of this notion of emergence in some wider interdisciplinary concepts (besides form and causality).

It is different for the problem of *epigenesis*, which seems connected but which in reality is very different. Under this term, effectively, there is firstly its historic opposition to the former idea of preformation in which an egg is seen as a miniature being which simply needs to be deployed. More generally, by epigenesis or epigenetics, we mean everything which, in particular throughout embryonic development, not only relates to genetic determinism, but is also controlled by other factors, which can cover a great variety of issues or processes (see section 1.2). For example, cellular differentiation varies locally depending on the site, whereas all the cells in an organism have the same genome, a recurring point that was put forward in the 1920s by T. Morgan in his key research into the genetics of drosophila. Let us add that the term "epigenetic" was outlined in morphogenesis by Waddington, by

---

73 In Andler, D., Fagot-Largeault, A. and Saint-Sernin, B. (2002). *Philosophie des sciences*, vol. 2. Gallimard, Paris, 939–1048.

attributing a very general theoretical meaning to it that expresses a plurality of potential development routes (denoted "chreodes"). This is the famous designation of an *epigenetic landscape*, including potential wells and potential barriers. With this idea, a choice of morphogenetic behaviors operates with a principle of optimality in response to the exercise of constraints that play the role of control variables.

In passing, it is useful to return to the thoughts of Stuart Mill (*Système de logique déductive et inductive*, 1843) who proposed a distinction between two types of laws. Certain laws, which he described as "homeopathic", correspond to a "composition of causes", in analogy with the vector addition of forces in mechanics. On the contrary, other laws, denoted "heteropathic", do not comply with this principle of vector composition (this would be the case for chemical reactions). If living organisms are very strictly composed of physical elements, their properties would result from these heteropathic laws, thus violating the principle of a simple composition of the properties of their constituents. Along the same lines, Broad referred to "intra-ordinal" laws for those that connect the elements of a particular organizational order (e.g. neurons) and "trans-ordinal" laws for relations between elements of different orders (in fact limited to between contiguous orders). We can observe in this an outline of the position held today by the interactions between elements in the same system or even of the multi-scalar approach.

While research into complex systems became the declared objective of certain laboratories, such as the Santa Fe Institute (United States of America), founded for this objective in 1984, in reality the need for a change of paradigm took shape much earlier and very progressively. Without revisiting the points of view of the biotheoreticians Rashevsky and Rosen, it is useful to note J. Bonner's comments on the subject in 1960, i.e. a little after the discovery of the structure of DNA in 1953. In an editorial entitled simply *Thoughts about biology*[74], Bonner was already indicating the need to take up a position at a level of abstraction above classical biology, by making the following statement of ambitious intent: "Biology is becoming a rigorous science with sophisticated laws and operational rules and theorems".

Since statements of this kind, which took a long time to be accepted, this ambitious objective has a history in itself, which we will talk about further on, concerning the panorama of mathematical tools in biology that we draw up (Chapter 5). For the moment, let us note that in its general meaning, the use of the formalism

---

74 Bonner, J. (1960). Thoughts about biology. *Amer. Inst. Biol. Sci. Bull.*, 10(5), 17.

for dynamic systems (differential equations) extended progressively in the following stages:

– a single global variable $y(t)$ as is seen in a univariate law of growth (population numbers);

– p interdependent variables $y(t)$, **y** vector;

– effect of position in a field **x**: $y(t, x)$ (local variations in activity);

– reaction + transport (diffusion, for example): models of reaction-diffusion which describe the process as spatio-temporal in nature;

– structured model: several classes of the same variable (stratification in classes of functional equivalence, age classes in a population, classes of cellular states);

– multi-agent model: several groups of individuals (biological associations);

– multi-scale model: consideration of several organizational levels (molecule, cell, tissue, organism, ecosystem).

3.12.4. Networks

According to a general definition, "a network is a set of objects that are interconnected and united by their exchanges of matter and information"[75]. We have already had the opportunity to refer to this formalism, having seen in particular the specific case of Petri nets. It is now necessary to situate it in a more general manner as a basic concept in the study of many biological phenomena, from cell biology to biology of populations and ecology.

In fact, this term designates a type of system organization which is based on the theory of graphs, more exactly in our words on directed graphs. The objective is to obtain a precise representation of the relations between the different objects or elements of the system. We therefore have a set of N nodes that feature the various elements, which are variously connected to each other by a number k of directed arcs which signify their relationships $\mathcal{R}$. Here, we will only look at non-valued connections. A network is a given combinatorics of the relationships between nodes, where each node is connected only to a precise number of other nodes. We observe that every directed arc that corresponds to the existence of a relationship in a given direction (source–target type: $x \rightarrow y$), i.e. $x \mathcal{R} y$, has the connotation of a direct causality, let us say local. But globally, of course, the principle of causality is presented as diffuse, distributed over the entire network, which provides a considerable extension of the principle of circular causality by T.A. Hutchinson (1948), which at the time completed the direct linear connection.

---

75 Parrochia, D. (1993). *Philosophie des réseaux*. PUF, Paris.

This principle of representation, relatively recent in its theoretical aspects, is currently an important theme in mathematics. It has universal scope, applying to a wide variety of systems that are encountered in both social sciences or economic sciences, and to natural sciences. This notion of a network is particularly relevant to biology, and is now considered to be an essential concept to understand living things. Thus, as significant examples, we can cite metabolic or genetic networks, neural networks or even ecological networks, all of which are communication networks (*lato sensu*) according to highly varied graphs that can be (but are not necessarily) of large dimensions and above all subject to variation (evolution of the arcs between nodes). The notion of a network in biology is both structural and functional in nature. Let us note in passing the significant difference with certain physical networks, like those used in crystallography, such as Bravais networks that were proposed in 1838 as a representation of the plant facies phyllotaxis (with well-known critiques from the botanist Plantefol concerning this strictly structural aspect considered *a priori* as purely hypothetical) (we will summarize this in Chapter 5).

Developed in the second half of the 20th Century, graph theory originated from a problem of mathematical recreation examined by L. Euler in 1735 and known under the name the "problem of the Seven Bridges of Königsberg". Königsberg (today known as Kaliningrad) is a city of Old Prussia, established on the two banks of a river and on two islands, and includes seven bridges that connect the four districts of the city. One of these bridges connects the two islands together, and the six others connect the islands to the two banks of the river. The following question then arose: is there an itinerary going from any point in the city which takes each of the seven bridges, and only once, before coming back to the starting point? The constraint applied to the graph construction is to use the various connections only once (here, it is a case of simple pathways, not directed graphs). L. Euler showed that there is no solution to this problem. Subsequently, it was necessary to wait until the 1990s to see the effective use of graphs that allow a natural network to be described or that allow one to experiment with the consequences of a variation of connections.

Mathematically, a graph G pertaining to a set or a system S is a sub-set of a Cartesian product of S by itself:

$$G \subset S \times S$$

which defines the nature of the relationships between nodes. That is, $x_i \, \Re \, x_j$ is the relationship between elements x_i and x_j.

Different types of networks need to be considered depending on the type of relationship between nodes[76]. One of the important points relates to the evolution possibilities of a given network. In effect, it is not sufficient to describe a system like this in a mathematical graph, because it is also important to specify the variation possibilities of the connections between nodes. This can be seen by means of an experimentation or simulation on the network itself. On this subject, let us note that various algorithms have been designed for the generation of networks (e.g. the Barabási–Albert model for the generation of scale-free networks).

Concerning the evolution of networks, we emphasize that the modifications of the connections within a natural system can have a sense of reaction or adaptation in response to this kind of constraint or stimulus (such as modifications of neural connections during learning or the failure of a biochemical site in a metabolic network). An important practical point is the *robustness* of a network, meaning its capacity to resist disturbances that can dismantle it. This property depends, of course, on the network topology, as well as on the probability of node deletion (existence of a threshold or a critical probability).

3.12.4.1. Random networks

Random networks are described by a distribution of the frequencies of the number of connections that follow the Poisson law of probability (a law that we will describe in Chapter 5, regarding random processes) (a bell curve of frequencies). In networks of this kind, there is therefore a small number of nodes that are either weakly or strongly connected, and a large number of nodes with an average degree of connection. It is useful to study the evolution of this type of network. Effectively, its size can be modified by making an experimental modification of the ratio of the number of nodes connected by arcs of a given length and the number of arcs present. This ratio has a threshold value beyond which there is a significant increase in the number of nodes, which can lead to a network of a very large size being created. This property of abrupt transition is observed in the phenomenon of percolation for which the mathematical model was introduced in 1957 (J. Hammersley). The well-known typical example is the movement of a fluid in a porous medium. This process has been studied using a network of specific sites that allow information (here the fluid) to be transmitted in given directions. Another situation is a set of sites that can fix a ligand with a given probability. This principle is reused in the study of the propagation of various evolving physical or biological processes (forest fires, for example).

---

76 Concerning the notion of network, refer to (Ricard 2008, Chapter 6) and (Zwirn 2006).

3.12.4.2. *Small-world network*

An important category of networks is the "small-world" category. Advocated by the American psychologist S. Milgram in 1967, this type of network is based on the following hypothesis: any individual can be connected to another via a short chain of connections. Milgram deduced the concept of "six degrees of separation" (using the average number of connections recorded) from his experience in psychosociology, studying the average path length for social networks. Subsequently, this question was studied by the physicists D.J. Watts and S.H. Strogatz, who proposed in 1998 the first model of the small-world network. Experimentally, on a regular network, they observed that the random addition of a few connections allows the direct length between two nodes to be reduced (from "very long" to "very short"). It appears that their work originated from observations of crickets' stridulations, whose behavior testified to a strong coordination between individuals over large distances.

Two properties characterize small-world networks *grosso modo*:

– the average distance between two nodes is proportional to the logarithm of the number of nodes (whereas in a mesh, the average distance is proportional to the number of nodes);

– the neighbors of a given node are often connected to each other (this does not take place in random networks).

3.12.4.3. *Scale-free networks*

A *scale-free network* is a network with an average degree of node connection k (number of arcs that connect a node to other nodes) that follows a power law: $p(k) = k^{-\gamma}$; $\gamma > 0$. A network of this kind has a limited number of highly connected nodes, and a large number that are weakly connected. The comparison between frequency curves that follow the Poisson law and those that follow this negative power function illustrates the difference with random networks. In terms of growth of a scale-free network, highly connected nodes tend to establish new links with other nodes, which weakly connected nodes cannot do (property of "preferential attachment"). In networks of this type, "hubs" are created through which various internode relations pass. Many networks are of this kind, for example the Internet, as well as metabolic networks[77]. These networks are considered to be resistance to the loss of a node or of an arc (e.g. inhibition of a metabolic function or of the regulation of a gene). Due to the network topology, accidents of this kind only modify the general behavior a little, hence the name *scale-free*, in which the dominant aspect is the importance of highly connected nodes. On the contrary, if these super-nodes are damaged, the network becomes vulnerable.

---

77 Albert, R. (2005). Scale-free networks in cell biology. *J. Cell Sci.* 118, 4947–4957.

Two examples of simple networks are indicated in Figures 3.12 and 3.13, based respectively on Goodwin's system of differential equations for genetic regulation and on qualitative modeling of logical kinetics by Thomas (see section 5.7).

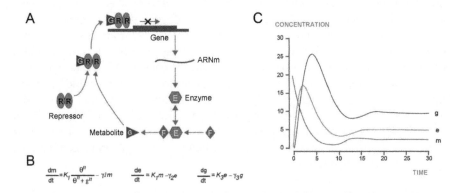

Figure 3.12. *Genetic regulation network based on Goodwin's differential formalism (indicated in B). C: result of a simulation (initial conditions and parameters set up in advance) that leads to a stable stationary state; the three curves correspond to the numbers of RNAm (m) molecules, of enzymatic protein (e) and of the metabolite G (corepressor g) (according to Thieffry and de Jong (2002)). For a color version of this figure, see www.iste.co.uk/buis/biology.zip*

The importance of the notion of network in biology has now been completely acquired in principle. It has been outlined by various authors, in particular S. Kauffman (*The Origin of Order*, 1993) who insisted on the property that certain complex systems have of exhibiting a spontaneous order. Thus, the order of a network configuration can emerge, meaning a parallel functioning (and not in sequence like how an algorithm would advance step by step) due to the fact that several nodes can be in activity at the same time. Enriched by the mathematical side of networks, this agrees with the principle of genetic determinants that is founded, for the execution of a given function, on the coordination of several genes (according to the notion of *operon*). It is also necessary to note that, in many situations, there are in fact networks of networks, such as metabolic networks. In the latter, the nodes are the metabolites and the arcs represent multi-enzymatic transformation reactions, where these are actually networks in themselves[78].

---

78 (Ricard 2008, Chapter 10).

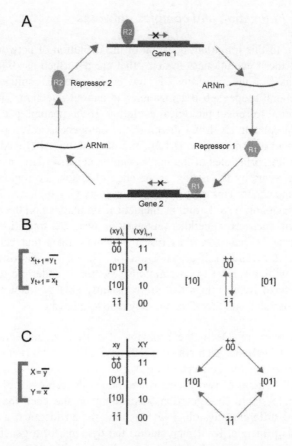

Figure 3.13. *Network associated with the Boolean modeling of logical kinetics of a system of two mutually exclusive genes. For a color version of this figure, see www.iste.co.uk/buis/biology.zip*

COMMENT ON FIGURE 3.13.– *B: synchronous functioning. The equations express the fact that at a given instant (t + 1), the coding gene for the first repressor (variable x) will only express itself if the second repressor (variable y) was previously absent (at instant t) (the same is true for the expression y). The system has two stable states [10] and [01] according to a cycle in which the two genes turn on and switch off in an exactly synchronous and indefinite manner. C: asynchronous functioning. The same stable states as in B, but the evolution depends, according to the states, on the commutation transition times (arrows), which is why one repressor can reach its action threshold more rapidly that the other (according to Thieffry and de Jong 2002).*

3.12.5. *Order, innovation and complex networks*

Let us return to this important question of the evolution of networks. Based on the behavior of these evolutionary systems that are described as *complex systems*, Kauffman points out the necessity of now being able to conjugate these two principles of general scope: self-organization and natural selection. This is in effect entirely insufficient to reflect biological evolution. If the genome is a network, and so dynamic at a cellular level and therefore the ontogenesis of an organ or of an organism, it cannot only depend on Darwinian selection, but additionally (and above all) on the founding principles of the representative network. This subject has been helped in recent years by the notion of the edge of chaos, the region of transition between order and chaos. This expression, proposed by C. Langton in 1990 and used in reality as a metaphor, results from mathematical simulations on the dynamic of an evolving system such as a cellular automaton, near its transition towards an unpredictable chaotic behavior. This research demonstrates that certain complex systems can tend towards self-organization (as shown by Kauffman in 1991). Biologists will willingly refer to an *adaptative system* to underline its connotation with the property of living things that are constantly able to adapt and to survive (dynamic homeostasis), and therefore possibly able to innovate.

Kauffman's work relates to the behavior of a Boolean network of N nodes (values of 0 or 1), such that each note depends on K other nodes (network NK). The basic idea is based on the concept of operon by Jacob and Monod. Thus, in the simplest case of a circuit of two genes, where one is the repressor of the other, we have two possible stable configurations depending on the gene that is activated, determining two different physiological functions for an identical genome. As an extension to this principle, Kauffman studies the dynamic of a genetic network in which each gene depends on K other genes. The transition from order to chaos depends on the relationship between the frequency of the numbers 1 and 0 in the whole network. The parameters P and K determine this transition. Whatever the value of K, there is a value of P for which the network has a non-chaotic dynamic (order side of the region).

This is the moment to describe the outlines of the current framework for different types of stability of a dynamic complex system. In addition to the three types of dynamic behaviors that are already well-documented in the context of differential systems: (1) stability (punctual or multiple by multistationarity depending on the initial conditions); (2) regular oscillatory regime (limit cycle); (3) instability by disordered and unpredictable oscillations (chaos), there is in effect a fourth type, which corresponds to the regime known as the edge of chaos. The effect of a disturbance is weak and relatively brief for the two initial cases (structural stability), significant and irreversible for the characterized chaos. Concerning the edge of chaos, the effect is in principle localized and long-term. Additionally, in contrast to other cases, it presents a capacity for innovation which has the meaning of self-organization. Observed in sociology and economics, this property is advocated by Kauffman for genetic networks. The simulation of its cellular automatons demonstrates that adaptive systems evolve towards this region of phase transition at the order–chaos border. The meaning of this behavior, in contrast to an ordered regime in the sense of classic stationarity of differential systems, is seen as an equilibrium or a compromise between stability and flexibility, which allows long-lasting innovation. However, this still remains a working hypothesis. The very useful contribution made by simulations using Boolean automatons that we have just drawn up does indeed require proof or validation that biological systems could themselves offer[79].

3.13. Game theory in biology

It was without doubt Buffon who, in relation to his thoughts on the "franc-carreau" game that we previously considered, insisted first on the role of psychology in the calculation of probabilities. He considered that in a random draw game, a kind of utility principle intervenes automatically in the player, meaning that their behavior is likely to influence the result. In other words, theoretical mathematical expectation is added to the question of behavioral strategy.

This is the objective of game theory, which is to formalize an *optimal strategy* among various possibilities or possible decisions. The mathematician E. Zermelo was the forerunner of this in 1913. His work was revised and developed by von Neumann and O. Morgenstern (1944). Initially attributed to the study of economic problems, this theory had initial applications in biology in the 1970s, with a view to modeling animal behavior, in particular in research by J. Maynard-Smith (1920–2004)[80]. This biologist was interested in the applications of mathematics in ecology and particularly in genetics (he was a student of J.B.S. Haldane), mainly in his relations with Darwinian evolution. In contrast to the economy, in which there is a certain liberty of decision, evolving processes are considered to be blind processes.

Let us draw up a representation of the principle of this approach in the problem of aggressiveness in animal populations, asking the question: does the course of evolution favor aggressive animals or is there a stable equilibrium that restricts the

---

79 Concerning this idea of self-organization and evolution at the edge of chaos, see the reservations that it provokes in ethology (Theraulaz, G., Spitzt, F. (dir.) (1997). *Auto-organisation et comportement.* Hermès, Paris).

80 Maynard-Smith, J. (1982). *Evolution and the Theory of Games.* Cambridge University Press, Cambridge. In 1968, this author published *Mathematical Ideas in Biology* with the Cambridge University Press, a small book (152 pages) that presents an outline of the way in which some basic biological processes are approached mathematically.

proportion of them? This is the dilemma known as "dove strategy versus hawk strategy". An aggressive individual (= hawk) intending to occupy territory or take power (e.g. a male with respect to females and in opposition with his competitors) puts his life in danger because he risks not transmitting his genes. Hence, from a teleonomic point of view (although this term is rejected), research focuses on the effect of behavioral strategies, such as flight or attempting to find a compromise, for example the approach rituals acting as a warning which can avoid a damaging confrontation. Maynard-Smith proposed a mathematical model based on game theory from which the concept of evolutionarily stable strategy (ESS) arose and has now become classic. We will provide an outline of this as a simple example of game theory in biology.

By means of some simplifying assumptions, we look for the gain G that results from the hawk F-dove Co game, in various meeting situations: (F, F), (Co, Co), (F, Co), (Co, F). The meeting noted (I, J) designates an individual that uses strategy I compared to an individual that adopts strategy J. The games matrix Γ groups together the average results of these encounters in terms of gains G and costs C:

$$\Gamma = \begin{bmatrix} \dfrac{G-C}{2} & G \\ 0 & \dfrac{G}{2} \end{bmatrix}$$

For example, the meeting between two hawks leads to a conflict with the average result of a shared gain G-cost C balance, each of them presumed to have the same probability of winning.

The gain of an average individual, meaning the probability of the adopted strategy, is a function of the proportions $x(t)$ and $y(t)$ of these two groups F and Co in the population:

$$\Delta = \begin{bmatrix} x & y \end{bmatrix} \Gamma \begin{bmatrix} x \\ y \end{bmatrix}$$

Finally, we presume that the choice of a strategy of this kind also depends on the difference (+ or −) between the resultant gain and the average gain of individuals in the population. Hence, the evolution equation of the group F:

$$\dot{x} = \frac{1}{2}x(1-x)(G-Cx)$$

This equation presents three possible stationary states. Other than the cases of exclusion of one or the other of the groups, there is an equilibrium solution known as a stable "polymorph" in which both groups co-exist. The proportion of hawks is then: $x^* = G/C$, an example of an evolutionarily stable strategy.

In this basic model, strategies are qualitatively set up *a priori*. An initial interesting extension consists of laying down the principle of an adaptation of the strategy depending on the meeting. This game model is known as *retaliatory*. The strategy R (*retaliator*) is added to the preceding strategies F and Co: a dove individual adopts a hawk strategy if they are subject to the aggression of a hawk individual, otherwise they remain a dove. The game matrix becomes:

$$\Gamma = \begin{bmatrix} v-u & 2v & v-u \\ 0 & v & v \\ v-u & v & v \end{bmatrix}$$

where $v = G/2$ and $u = C/2$. Taking into account the condition of normalization of the numbers in the three groups $x(t) = y(t) = z(t) = 1$, we have the dynamic system:

$$dx/dt = x\left(-u + 2ux + (u+v)y - 2uxy - ux^2\right)$$
$$dy/dt = xy(2u - v - ux - 2uy)$$

One of the solutions (depending on the values of u and of v, with $v < u$) can (depending on the initial conditions) be the stable equilibrium $(x^*, y^*, 0)$ of the coexistence between the hawk and the dove. However, in their initial proportions, an elimination can take place instead of the hawk strategy, since the population ends up containing only the dove and "retaliator" types (Figure 3.14).

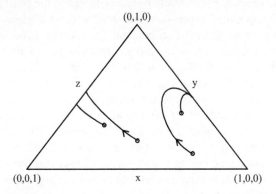

Figure 3.14. *Hawk-dove-retaliator game. u = 2; v = 1*

Another basic game is "rock-paper-scissors". These illustrative terms signify three types of strategy with the results: (i) the scissor wins over the paper but is blunted by the rock (and so becomes inert); and (ii) the paper wins over the rock because it can cover and therefore paralyze it. The matrix of gains is given as:

$$\Gamma = \begin{bmatrix} 0 & 1 & -1 \\ -1 & 0 & 1 \\ 1 & -1 & 0 \end{bmatrix}$$

We demonstrate that the trajectories representing the variations in quantities $x(t)$, $y(t)$, $z(t)$ are closed curves around a center of equilibrium (1/3, 1/3, 1/3). This game then leads to regular oscillations in the proportions of individuals that play this strategy *a priori* R, C or P. In this case, characteristics of oscillations (amplitude, period) depend on the initial state. The introduction of a parameter that modifies the previous behavior can determine a bifurcation of the dynamic, which leads either to the installation of a unique closed trajectory (limit cycle) (i.e. regular oscillations), or to a damped oscillatory regime that leads in the end to a unique (punctual) stable state.

Other more elaborate games need to be pointed out, such as games with two gains matrices Γ that correspond to situations in which two groups co-exist, each with their own strategy (e.g. males and females), or even games with a cooperation strategy (on the basis of the classic game known as the prisoner's dilemma, based on cooperation by admitting and/or denouncing vs. denying). The latter leads to an altruistic effect, where the gains can increase due to cooperation between players.

This issue of behavioral strategy applied to evolutionary genetics results in a new vision of *natural selection* by Darwin, consisting of looking at the gene level and not at the level of the species, which resulted in the concept of the selfish gene (Dawkins in 1976). This metaphorical expression, presented without a teleonomic *a priori*, is not contradictory to an altruistic effect and its starting point is moreover the prisoner's dilemma game, where a cooperation strategy can be equivalent to an appearance of altruism. It simply means that if this gene dominates, it is "as if" the selection was favoring its expansion, which would be determined by its behavior towards others. A debate followed, on the one hand, concerning the pertinence of the "gene level" in question and, on the other hand, concerning the facts of conflict with altruistic behaviors that benefit the group or the species. Regardless of the case, game theory has taken up an important place in genetics, where we consider that current genes are those that, during evolution, have benefitted from the best possible strategy, a condition of their expansion at the scale of the population of the genomes determined at their conception.

In a wider sense, beyond this particular branch of genetics, a new field has been created, known as **evolutionary biology**. Its objective is the analysis of population dynamics whose development results, other than the classic state variables of the state of the system, from an additional determinant designated under the generic term "strategy".

Moreover, this term is well-known in biology in the designation of two types of population growth, animals and plants, depending on whether the reproduction rate (increase and precociousness of the fecundity or vegetative multiplication) or the capacity of the environment in terms of available resources (increase in individual biomass and lifetime) is favored. Beyond the first research works that were above all aimed at research into food, these two types of demographic strategy in a fluctuating environment are designated (Pianka in 1966) by the terms "r-strategy" and "K-strategy", in reference to the parameters r and K of the logistical growth model that we have mentioned. Returning to this idea, game theory consists of taking into account the fundamental property of living things which is the *ability to adapt to the environment*.

3.14. Artificial life

The term "artificial life" was proposed by C. Langton around 1980 to designate the application on an artificial medium of the principles or laws of living systems. This artificial medium can either be abstract in nature, like a grid or a lattice that represents a field of living cells in development on a given substrate, or physical in nature, in the same way as everything that relates to the construction of biomimetic automatons that simulate the behavior of a superior animal.

Although the name "artificial life" can be applied to all work, even simply calculational or graphical (e.g. the simulation of plant architectures) which results in reproduction of these properties of a living system (i.e. a kind of synthetic biology where our opinion on the pertinence of this designation is not important), which is used in particular for everything that relates to the theory of automatons.

We have seen that the origin of this theory lies in a set of research works, in particular those of von Neumann in the 1950s–1960s and those inspired by cybernetics such as the science of regulation of a system (Wiener in 1947). Let us summarize the connection that can be established between artificial life and biology itself.

3.14.1. *Biomimetic automatons*

Mankind has always attempted to construct machines inspired by living things which simulate their effects or behavior[81]. A long time before the developments in robotics that we are currently experiencing, various automatons were created. Particularly well-known works are those of Leonardo da Vinci during the Renaissance, then those of J. de Vaucanson in the 18th Century, with his numerous and famous mechanical constructions. Vaucanson's[82] duck is often mentioned, an automaton with 400 parts that had various functions (breathing, digestion, excretion, locomotion, but not reproduction). However, there is no biological aspect to the design of these machines. On the contrary, with the advent of *cybernetic automatons* (the "electronic animals" of the physiologist G. Walter and the engineer A. Ducrocq, the Homeostat of Ashby), a step forward was made, because they were founded on notions of regulation and control that we knew were common to living things and to electronic machines. In fact, the creation of cybernetic machines was the preliminary to the project for artificial life.

Regarding this analogy between living organisms and machines, we often consider, along with Jacob, that "animal and machine, each of the systems then becomes a model for the other [...] animals can be described in light of machines. Organs, cells and molecules are then united by a communications network"[83]. In reality, physiology preceded cybernetics with the demonstration of hormonal regulations determined by communication between the production site (endocrine gland) and the reception site. Discovery of the first hormone, secretin, was made in 1902 by W. Bayliss and E. Starling. Of course, the role of cybernetics was essential in generalizing and reinforcing the importance of the notion of communication in all systems at the same time as directional diversity and the sign of connections (*feedback, feedforward*).

A new objective then appeared with the issue of *learning*. The importance of learning processes is already recognized in living things where adaptation is a condition for survival, and also in all machines which target a performance with respect to an assigned objective, namely prefiguration of the robot. Thus, in this field that relates to our subject, a connection is established between these "bionic machines" which originate from artificial intelligence methods and animal psychophysiology, both of which attribute great importance to mathematics and

---

81 The idea of a material automaton seems to have taken shape very early on, in the 4th Century B.C., through the philosopher and mathematician Archytas who created a bird that was capable of flying, a precedent to Leonardo da Vinci and Jacques de Vaucanson.
82 A diagram is provided of this by Giavitto, J.-L., Spicher, A. (2006). Morphogenèse informatique. In *Morphogenèse. L'origine des formes*, Bourgine, P., Lesne, A. (dir.). Belin, Paris, 328.
83 Jacob, F. (1970). *La Logique du vivant*. Gallimard, Paris, 273–274.

computer science and which can moreover be considered more technical than conceptual. This aspect draws up a relationship between this research and what is commonly known as bioinformatics.

3.14.2. *Psychophysiology and mathematics: controls on learning*

At the interface between physiological mechanisms and cognitive sciences, a significant contribution has been made by mathematics and subsequently developed, meaning that of **networks of formal neurons** (or **artificial neurons**).

By "network of formal neurons", we mean a structured set of numerous interconnected units of calculation that work in parallel and have connections that are likely to evolve. A connectionist network of this kind is capable of learning to recognize and class forms, by means of learning in advance from examples proposed by the operator with a view to a certain objective or a project to implement.

The first mathematical model of a neuron was created by the neurologists W. McCulloch and W. Pitts in 1943, designed as an analogical tool to analyze relations that may exist between a computerized calculation algorithm and the function of the human brain. From this basic model, other types of formal neurons were developed. In the same way as dendrites and the axon of a biological neuron, a formal neuron is a mathematical object that has several inputs and an output, and that has a learning function that modifies internal connections. Each input is weighted by a "synaptic weight", and the output is determined by an activation function. This, in the model by McCulloch and Pitts, is the Heaviside function which takes the values 0 or 1 (step function) (this is a primitive version of the Dirac distribution). Other models use a less abrupt function, such as the logistics function or the hyperbolic tangent function. Due to the calculation errors that are a necessary part of learning, a second derivative function is used. A valuable extension, which greatly increases the calculation power, was designed in 1957 by F. Rosenblatt under the name of "*multilayer perception*". It consists of a set of layers of neuron networks, connected between themselves in a given direction from the input layer to the output layer.

This research is within the realm of what is known as "*artificial intelligence*". An initial development of this relates to what we call expert systems. By "expert systems", we mean software that responds to a certain number of questions in order to then act as a *decision tool*. It functions with a set of varied information (knowledge basis) and logical rules (reasoning). As examples of application, we can mention diagnostics tools in medicine and reasoned management of a crop in agronomy (choice of inputs, taking into account the nature of the soil, the climate, etc.). For a number of years, research has focused on the question of *deep learning*, a term that combines powerful computer methods of automatic learning, using in particular large networks of multilayered artificial neurons. These methods are fast-growing, aimed, for example, at the problems of recognition of forms (images, faces, texts) (see section 3.7) or voice recognition (without mentioning the confrontations in the games between humans and computers that have received so much media attention).

Here, let us look in detail at the development of this analogy of calculation in this particular field of *psychophysiology*. A new term was created from this: "animate". In this term that combines both the animal and the material, we designate any artificial entity that is not purely abstract, that has a high calculation capacity and an ability to adapt, and which is able to simulate a biological behavior (higher animals) (Wilson in 1985). We thus evolve towards another field, robotics, where animates associate mechanics (movement), electronics (regulation) and calculation (recognition and decision). The aspect that is of interest to us here is that these artificial systems possess a property of self-organization, where the connections of formal neurons are subject to reorganizations brought about by learning. Regarding the use of this term in biology, it is necessary to specify that here, this consists of self-organization in a weak sense according to the distinction made by Atlan. In other words, there is a programming of general, non-specific learning rules, and the meaning of functioning is defined *a priori* by the designer.

In summary, on the basis of this fundamental idea that a bionic system is necessarily capable of adapting its operation to different situations, we encounter one of the characteristics of the history of biology, namely the tendency to reject or at least minimize everything relating to a strict predetermination or a preformation, however without allowing ourselves to exclude these points of view, which are in part true. This tendency or position of principle therefore promotes the idea of natural selection and widens its scope. Leaving behind its initial field of the evolution of natural living things, it is at the origin of *genetic* (or evolutionist) *algorithms* whose field of application widely surpasses that of biology.

3.15. Bioinformatics

Under this name, as is heard most often in practice, it is not at all a case of grouping together all applications of computer science in biology, but instead of the well-defined field of investigation of the interpretation of genetic information.

Based on the analysis of DNA sequences, it has various objectives, from molecular modeling to taxonomy (molecular phylogenics). This biology has the particular characteristic of requiring the processing of enormous quantities of data and the use of gene banks for comparison, requiring the use of appropriate computer science algorithms. Due to this necessarily technical aspect, in which correlations or similarities are sought rather than relationships of causality, bioinformatics is located a little outside the scope of our subject, which we dedicate to a review of the connections between mathematical concepts and living phenomena that mark the development of biology. Due to this, we simply note here the important position that it holds in contemporary research.

More generally, it is also necessary to think about the position held by computer science given the widespread use of computers. Since by means of its calculation power and graphical imagery, this allows processes to be simulated, it sometimes takes precedence in the work of a mathematization, meaning help in understanding a phenomenon through its properties and not only through its phenomenological reconstruction. This explains the recognized reticence of some biologists and mathematicians and not to mention philosophers with regard to what they see as a risk of misappropriation, whereas this can and must be an original aid that is well defined by its objective, rather than a competitor that is sometimes disconcerting. A simple but very illustrative example is brought to us with the case of multidimensional statistical analyses (factor analyses) that are very widely used to process large files of data, in particular in ecology and social sciences. Imposing the limit of this exploratory objective, however valuable it is, means forgetting to look for what these methods can demonstrate, meaning the existence of a *latent structure* whose interpretation needs to be provided concerning the biological nature of factors that are outlined in this way[84]. However, of course, everything depends on the objective or the motivation involved when biologists use a particular mathematical or computer science tool... Finally, let us recall this other type of simulation, currently well-developed, and that we have already talked about, namely the construction in computing of plant architectures, whether this is based on the principle of formal grammar that generates all kinds of morphogenesis, or that of random processes of growth and branching.

---

84 This question is discussed in Buis, R. (2016). *Biomathématiques de la croissance*, Chapter 21. EDP Sciences, Les Ulis.

4
Laws and Models in Biology

We have previously mentioned the terms "law", "model" and "theory" several times. Without detailing the epistemological distinctions concerning them[1], we need to specify the way in which biology works and what it tends to do when it speaks of "law" or "model", in a mathematized form, whether for the simple technical purpose of representation or to claim to see in it some prolegomena that are still embryonic in a theory to come.

The term "law" is widely used in different disciplines with the notable exception of mathematics, where its use is limited to a few well-specified areas. This is the case of the "laws of probability" fixing the distribution of a random variable (density law and its integral or distribution law) and that of the "laws of algebraic structure" (laws of internal and external compositions, morphisms). Elsewhere, we do not speak of "laws" but of "lemmas", theorems or corollaries that we carefully distinguish from simple conjectures, which are only proposals not yet proven. Thus, we do not say "law of the square of the hypotenuse". As F. Gonseth remarks, "where everything is law, we no longer mention the word law"[2]. Why the thing and not the word? We can think that this reluctance of language is linked to the idea of cause, underlying the laws of physical phenomena, but considered absent from the mind of the mathematician who navigates a world of abstraction (without not also being interested in its applications). It is nevertheless clear that "the mathematical language intends to be that of causal rigor"[3].

---

[1] On this recurring theme in philosophy of science, see, among others, the works by Andler *et al.* or Varenne, previously cited.
[2] Gonseth, F. (1934). *Science et Loi. 5th International Synthesis Week*. Alcan, Paris, 12 (cited by Delsol, M. (1985). *Cause, Loi, Hasard en Biologie*. Vrin, Paris, 81).
[3] Bruter, C.P. (1996). *Comprendre les mathématiques*. Éditions Odile Jacob, Paris, 287.

In experimental sciences, the term "law" generally refers to any repeatable relationship, either between variables or measurable characteristics or between processes, and having a certain degree of generality that associates a defined domain of validity with it. Before looking at the situation in biology, it is worth clarifying a little bit how physics and chemistry, with their arsenal of laws expressed in mathematical language (which is their preferred language), conceive this term[4].

First of all, it may be noted that the terms "law" and "principle" are not always very well distinguished in their use. We speak unambiguously of "principle" to designate a fundamental basis in scientific discourse, of a nature quite similar to what an axiom or postulate is in mathematics. In mechanics, for example, we have the principle of equality of action and reaction and the principle of inertia. We can also mention the principles of mass and energy conservation. Thermodynamics, on the other hand, is based on its two principles that introduce the concept of entropy. The various physical laws respect such basic principles. However, there are cases where principle and law are used interchangeably. This is the case for Newton's laws, in particular, his second law expressing the mathematical relationship between force, mass and acceleration, a fundamental relationship that is described as both "principle" and "law". In biology, the term "principle" is sometimes favored by some authors, as in the 19th Century with É. Geoffroy Saint-Hilaire. Specifically, he set forth three "principles" governing the morphology of animal organisms, which are considered as invariants: the principle of unity of plan or organic composition, the principle of connections and the principle of organ balancing. For his contemporary F. Cuvier, it is a question of both principle and law to underline the importance of organic correlations.

How to define a law? We can agree on the simple and ideal position of C. Bernard, for whom a law is none other than the clearly explained relationship of an observed variable to its cause, according to its well-known purpose: "the law gives us the numerical ratio of effect to cause, and this is the goal at which Science stops". It is understood that these words are to be taken in the plural, since any observation can result from several elementary causes, and any cause can affect several processes. Substituting cause by factor, we will speak of "multifactorial determinism". On this reference to the principle of causality, we can use the words of Montesquieu, who, in another field but undoubtedly marked by his own studies and scientific experiments, said that the law is the "necessary relationship that derives from the nature of things".

Let us remain on these generalities[5] for a moment to specify that *laws are conditioned*, i.e. their validity is always related to a specific domain setting the range

---

4 Buis, R. (1994). Lois et modèles en biologie. *Trans-disciplines*, 8, 8–16.
5 See Ullmo, J. (1969). *La Pensée scientifique moderne*. Flammarion, Paris.

of magnitude of the variables studied. One of the examples often cited is that of classical mechanics, for which Newton's laws are only valid for very low speeds in relation to the speed of light, or the case of geometric optics, for which the laws cease to apply for distances close to the wavelength in question.

It follows that divergence from a given law either leads to the establishment of a more general law (by changing the formalism or by introducing other determinants) or is a sign of its limitation to the domain of values considered until then. We know the famous case of the Boyle–Mariotte law, linking pressure and volume of a gas: $PV = Cte$. Established for a given temperature and a relatively low pressure, it is generalized by the law of perfect gases that involves temperature T: $PV = RT$, where R is the constant of the perfect gases. J.C. Maxwell has clearly shown that this law illustrates the duality of *macroscopic* (phenomenological) *law*/*microscopic law* based on a statistical approach at the particle level (velocity distribution of collision-prone molecules). Under some assumptions that the pressure results from the collision of molecules and that the temperature is equivalent to the average kinetic energy, J.C. Maxwell finds the empirically established Boyle–Mariotte macroscopic law. Its demonstration establishes the introduction of the calculation of probabilities in the description of physical phenomena using laws duly specified in their scope of application, a step also marked in thermodynamics by L. Boltzmann with the notion of entropy.

In physics, A. Einstein distinguished between "integral law" and "differential law". This simple formal distinction is well illustrated by the mechanical example of movement. Kepler's laws (planetary motion) are of the first type: they express the result of an action. Newton's second law, by its differential form, refers to the "how" by using the concept of force as the cause of motion. It focuses on the variation of the system in question (elementary variation per unit of time). This implies specifying the nature of the force invoked. For example: gravitational force meaning an interaction between two bodies. In short, it is a question of distinguishing *descriptive law* from *causal law*. This distinction is hardly operational in biology, first because it often speaks of laws outside any mathematical formalization and especially because it can only rarely refer to a well-established causality, at least in a simple and condensed way. It follows that in biology, the term "model" is used (as a system of interactive equations), which would be more appropriate than that of "law". With these general remarks in mind, let us now consider some specific examples of laws, first those expressed in vernacular language (which have their own interests) and then those, more elaborate, that take the form of mathematical relationships.

4.1. Biological laws in literary language[6]

The following two examples, to which we limit ourselves, illustrate well this kind of biological law, the degree of generality to which they claim at the same time as their qualitative empirical basis (which does not exclude their interest).

4.1.1. *The law of Cuvier's organic correlations (1825)*

"All being organized forms a whole, a single and closed system, whose parts correspond to each other, and contribute to the same final action through a reciprocal reaction. None of its parts can change without the others changing also, and therefore each of them, taken separately, indicates and gives all the others."

This holistically inspired law accounts for the physiological point of view at the level of the organism, the form being linked to functioning. It prefigures the objective of a systems biology.

Associated with this law as its corollary, F. Cuvier sets out a principle of hierarchy, *the principle of subordination of characteristics*, from which it follows that it becomes possible to reconstitute the entire organism if some of its parts are available. "He who rationally possessed the laws of organic economics could remake the whole animal". Therefore, the foundation of these new disciplines was established: comparative anatomy and paleontology.

4.1.2. *The fundamental biogenetic law*

This very general term also corresponds to different statements such as *von Baer's law* (1828) and *Haeckel's law* (1868, 1884).

Haeckel's law states that "the embryonic development of an animal species reproduces the different stages passed through by its ancestors during the evolution of the species", or "ontogenesis is a brief summary of phylogenesis". The well-known examples are the development of the respiratory system in vertebrates, with the metamorphoses of amphibians (from the embryo to the terrestrial adult,

---

[6] Le Guyader, H. (1985). Les langages de la théorie en biologie. Doctoral thesis, Université de Rouen, Rouen; (1988). *Les Langages de la théorie en biologie*. Vrin, Paris, Chapter 3, from which we borrow the quotations in this section.

passing through the aquatic form of the tadpole), or the gill slits of the mammalian embryo, a transitional state, a "memory" of an ancestral disposition[7].

Contrary to E. Haeckel's conception of the increasingly early appearance of the adult stages of ancestors, von Bauer's law considers that "the appearance of general characters precedes, during the ontogeny of a given species, that of special characters".

It can be noted with G. Canguilhem that in E. Haeckel's case, this biogenetic law "is less the inductive conclusion of a research than the guiding principle of a universal system"[8]. It is therefore not in vain that this "law" is referred to as "recapitulation theory".

Another example of the link between law and theory is given with the interpretation of "young forms of leaves" in botany. Since the shape of the leaf blade can vary greatly during the ontogenesis of a plant, the so-called "youthful" leaves normally appear (without any trauma) at the beginning of ontogenesis.

According to H. Gaussen's "theory of pseudo-cyclical evolution or over-evolution", youthful leaves are a primitive type corresponding to a return to an ancestral form, showing the link between ontogenesis and phylogenesis (see Y. de Ferré on gymnosperms).

NOTE.– On this question, we will simply note that in general "a law reflects a set of observations, while a theory reflects a set of laws". On the other hand, "a theory applies to several phenomena"[9]. For example, the Newtonian theory of universal attraction, with a well-marked validity domain, explains Kepler's laws (planetary motion) as well as Galileo's laws on the movement of an object on an inclined plane or that of the pendulum.

4.2. Biological laws in mathematical language

Such laws are rare in biology, incomparable to the number and diversity of laws in the physical sciences. The most remarkable are undoubtedly the laws of formal genetics (Mendel, Hardy–Weinberg), first because of the simplicity of

---

7 Gould, S.J. (1977). *Ontogeny and Phylogeny*. Harvard University Press, Cambridge, 66–68.
8 Canguilhem, G. *et al.* (1962). *Du développement à l'évolution au XIXe siècle*. PUF, Paris, 39.
9 Holland, J.H. *et al.* (1986). *Induction: processes of inference, learning and discovery*. MIT Press, Cambridge.

their formalism and also because they are based on the concept of the gene under its two allele states (dominant/recessive), a concept that was completely abstract at the time when these laws were proposed. Two main reasons explain this situation of relative rarity of these laws in biology.

On the one hand, because of the variability of biological observations previously highlighted, many relationships are statistical in nature, combining in the same equation a deterministic part (the law itself) and a random part (reflecting the intrinsic variability of the observations). We have seen this with regard to experimental designs validated by analysis of variance. We will discuss this again with the formalism of random processes (Chapter 5).

On the other hand, this other characteristic of biological processes, that of being the work of systems, leads to the need to pose problems via a set of relationships, rather than to focus on the simplicity that laws with a limited number of variables themselves carry. Therefore, biology cannot mention a corpus of elementary macroscopic laws that are specific to it and as simple as, for example, Ohm's law in electrokinetics or Snell–Descartes' law in geometric optics (refraction).

Let us illustrate this with the example of the formalization of a growth process. The biologist has a kind of catalog of growth laws, which is still being extended in order to improve their suitability for observations (see their review in Buis (2016). Many of these laws result from changes in a small number of basic equations (such as exponential or logistics). However, these are often minor modifications, which are more a simple addition of parameters than a call to new basic quantities or assumptions. Without ignoring their possible practical interest for a phenomenological description of a particular phenomenon, we know that a more satisfactory solution consists of a change of reference frame, i.e. dimensions, from a single growth equation to *a set of interactive differential equations*. For example, in the formulation of variations in the N population size of a population, it is the transition from a simple global law *N(t)* to a structured model based on the existence of several categories within that population. It is the consideration of several states of growth (such as those defined by the age structure of the population) and therefore of several variables, instead of reducing the problem to the behavior of a single global variable, that would be sufficient to describe the phenomenon. What seems trivial in demography is not always so for many growth phenomena where the number and nature of state classes can remain problematic.

With regard to these two remarks, let us develop a little bit this distinction between statistical or empirical laws and theoretical laws.

4.2.1. *Statistical laws*

Here, we empirically search by induction for a mathematical relationship between a variable called "explained variable" and one or more observed variables called "explanatory variables" or predictors, according to the general formalism of a multiple regression:

$$y = f\left(x_1, x_2, ..., x_j, ..., x_p\right)$$

For example, the classical methodology of multiple linear regression (*f* being a first-degree polynomial with separate variables without interaction) easily allows the best predictors *xj* to be chosen from a set of *p* variables assumed *a priori* to explain *y*. The term "explain" has a simple statistical connotation of correlation, not causality. For this purpose, appropriate software of the "stepwise regression" type with analysis of variance at each calculation step is available, allowing only a number $m < p$ of "statistically useful" predictors to be retained.

Another multivariate formalism is the so-called "multidimensional data analysis" In principle, quite different from multiple regression, where we distinguish between explained variables and explanatory variables, the aim here is to treat *p*-correlated variables on an equal footing, whose dependence on underlying variables called "endogenous" or "latent" variables, factors or components is sought. This is the distinction between regression and correlation. The standard method is *factorial analysis in principal components*. Developed by H. Hotelling in 1933 and originally confined to the analysis of psychometric test results, this methodology has seen considerable development of its applications in various fields using computer technology, particularly for analyzing large data files (e.g. in ecology or the human sciences). From this type of correlation analysis, we do not, strictly speaking, draw explicit laws linking the variables considered. Exploratory in nature (at least at the beginning of the work), the originality of factorial analyses is to highlight the existence of groups of variables on the one hand and, on the other hand, the antagonisms of some of these groups. It should be noted that these methods basically consist of searching for a latent structure, i.e. detecting the relationships between observed variables and factors. It is these *relationships between the observed and the hidden underlying* that can have the value of laws, at least to the extent that it is possible to give a biological interpretation to these latent variables.

4.2.1.1. *The law of allometry*

As an example of a statistical law, let us take the law of allometry, widely used in biology to numerically express the relationship between two variables. Originally, it was the relationship between two dimensions of an organ or an organism, or the relationship between the mass of an organ and the total mass of an organism. It is a first approach for studying the shape and differential growth of the different parts of a whole.

This was initially the purpose of Huxley–Teissier's law (1932–1934)[10] according to a power function linking the variables y_1 and y_2:

$$y_2 = by_1^a \text{ or } ln(y_2) = ln(b) + a\, ln(y_1) \qquad [4.1]$$

Parameters *a* and *b* are referred to as *allometric scaling* factor and proportionality or normalization coefficient, respectively. Thanks to this linearization in log–log graph, it is possible to highlight and test by statistical regression the existence of an allometry line. The principle of this relationship was applied to a number of situations depending on the species and variables, which can be both physiological (such as metabolic indicators) and morphological (dimensions or biomasses).

The success of this law is explained by the possibility, when this relationship is verified, of highlighting either the existence of a single allometric line whose parameters *a* and *b* characterize the species (or group of species) and the character pair, or on the contrary the existence of a partition of the value domain. In quite a few cases, we can indeed have not a single line in log–log coordinates, but two or more lines corresponding to a plurality of relationships according to the order of magnitude of the variables. This is referred to as slope rupture or discontinuity. A value of $a \neq 1$ indicates the fact of a differential growth of these two variables (heterogony or growth disharmony). If $a = 1$, there is isometry.

Different types of allometry should be considered depending on whether one is interested in a set of species or phyla (phyletic allometry), or the same species at a given stage (size allometry) or during its development (growth allometry or ontogenetic allometry). It is therefore either a static study or a dynamic study.

---

10 The names of J. Huxley and G. Teissier are generally associated with each other; their work was almost concomitant and their terminology was clarified in a common note published in 1936 jointly in *Nature* and in *C.R. Soc. Biol. Fr.*

The latter type of situation is well illustrated by the relationship between the mass of certain organs and the total biomass of the individual over the course of mammalian development. For some organs, the constancy of their relationship is observed, while for others, there is a clear discontinuity in their covariation, which generally coincides with a specific ontogenetic stage. In this type of case, it is noted that this change in the law in the interdependence of the two variables, i.e. their differential growth, is of a physiological hormonal nature (Figure 4.1). Many other examples are known, such as in plants where the change in the relationship between root organ biomass and aerial organ biomass is related to the sexualization of the apical meristem (floral induction) (Figure 4.2). Another example is a change in environment during animal ontogeny, such as the change in pressure during the transition from embryonic and fetal condition to postnatal growth (Figure 4.3).

Dynamic allometry is of particular interest to us because of its relationship to the growth law of each of the variables. From their first works, J. Huxley and G. Teissier showed that their relationship could simply be explained by the proportionality of the specific growth rates of the two variables,

$$\frac{1}{y_2}\frac{dy_2}{dt} = k\frac{1}{y_1}\frac{dy_1}{dt},$$ [4.2]

under the following two assumptions: (i) proportionality of each of these growths to the consumption rate of the same nutrient *pool*; (ii) constancy of nutrition/growth efficiency during growth.

This important issue of a connection between this allometric law and the growth law of each of the covariate variables is often overlooked because, for most published observations, sampling is not based on the homogeneity of growth states. However, it is easy to show that the Huxley–Teissier relationship is only verified in certain cases, particularly when the two variables are in a phase of exponential growth during which the specific velocity is constant:

$$\frac{1}{y_i}\frac{dy_i}{dt} = Cte.$$

However, this only corresponds to a part of most growth curves. In particular, the famous logistic law of growth, a classic function for sigmoid curves, can only allow the Huxley–Teissier relationship for the approximately exponential phase of the beginning of growth. We show that over the entire logistics curriculum, the allometry relationship is, in fact, a hyperbolic function and not a power function.

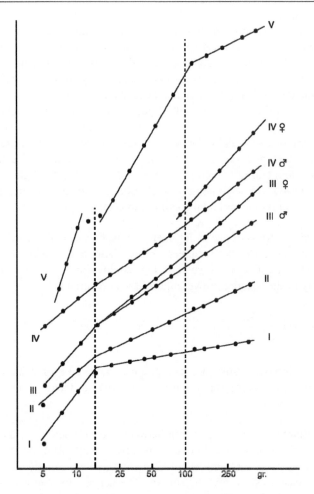

Figure 4.1. *Growth allometry in rats: mass variations of various organs (in ordinates) as a function of the total mass of the organism (in abscissa) (based on (Teissier 1937))*

$$y_2 = \frac{K_2 y_1}{\alpha K_1 + (1-\alpha) y_1}$$

where K_1 and K_2 are the limit values of these two growths and α is a parameter depending on the relative position of the inflection points of the two sigmoids. In this case, the power function [4.1] is inaccurate. The same is true for other usual growth laws (such as the Gompertz function), each with its own allometric relationship. Huxley–Teissier's relationship is therefore not a general law. Given its frequency of use, it should at least be presented as a simple approximation that can be used on only one growth phase.

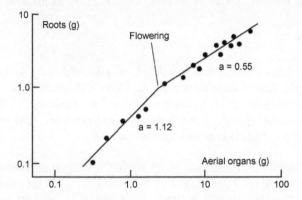

Figure 4.2. *Weight growth allometry between underground organs and aerial organs during developments of the grass Lolium multiflorum (based on Troughton (1960))*

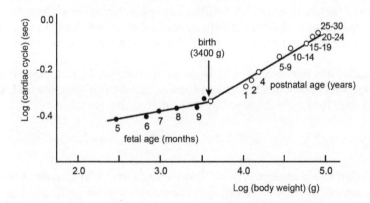

Figure 4.3. *Physiological allometry: relationship between the cardiac cycle and body mass in humans. •: embryonic growth; o: postnatal growth (based on Günther et al. (2003))*

4.2.1.2. *Physiological or metabolic allometry*

The principle of an allometric relationship is of particular importance for physiological variables that can be used as metabolic indicators, such as heart rate or caloric exchanges in animal physiology. Long before J. Huxley and G. Teissier, it was F. Sarrus and J.F. Rameaux who first studied the question in 1839, followed by M. Rubner (in 1883, 1908) with his famous experiments on heat production in dogs.

The experiments of these authors led to what is called the *"law of surfaces"* on the proportionality between the energy consumption measured by caloric loss Q per unit of time and the surface area of the organism, i.e. 2/3 of its mass M:

$$Q = k M^{2/3} \qquad [4.3]$$

Of course, this empirical law is related, in principle, only to homeothermic animals and under basic metabolic conditions. Under this condition, it is particularly true when comparing oxygen consumption (respiratory frequency × volume) in two animals of the same species, from which a correspondence between geometric similarity and physiological similarity can be deduced.

In terms of *dimensional analysis*, which is often used, heat production per unit of time has the dimension of a power (energy/time): $M L^2 T^{-3}$. On the other hand, if we refer to the theory of similarity in biology, developed by L. Lambert and G. Teissier (1927), who *a priori* posed the dimensional equivalence between the two independent dimensional quantities T and L, it shows that there is indeed proportionality between heat production and body surface area according to the relationship [4.3]. However, this presupposes that this total area itself is proportional to the actual radiant area, making it difficult to apply this law of surfaces to a set of different species or to different growth stages in their morphology. In other words, this law of surfaces has the meaning of a relationship of geometric similarity.

In reality, this relationship [4.3] is far from universal. In fact, other empirical relationships are known that reflect many and varied observations quite well, such as *Kleiber's law* (1961), which adopts the value of exponent $a = 3/4$:

$$Q = k' M^{3/4} \qquad [4.4]$$

This relationship remains very empirical without any solid theoretical basis. Let us give some illustrations. For example, the power function [4.4] is valid, among others, in small aquatic organisms and unicellular organisms, i.e. under physiological conditions where hydrostatic thrust modifies the effect of gravity. This condition is also found in embryonic and fetal development *in utero* (see Figure 4.3). In another situation, this exponent value 3/4 applies to very different species, such as large mammals (cattle), which Brody (1945) extensively studied for growth energy, while it is the value 2/3 that applies to small mammals. On this subject, let us note as quite different the search for a height allometry on species of varied body mass. This is the case of the application of the relationship [4.1] to a very diversified set of mammals whose adult mass varies from a few grams (mouse) to several tons (elephant) in order to verify the constancy of the metabolic rate by relating the allometric relationship to the individual mass. It should also be noted that this exponent 3/4 applies to physiological variables of a completely different nature, such as the xylemic flow in plants (transport of crude sap) compared to individual biomass.

This extreme diversity of observations shows that the value of this type of law is more phenomenological than explanatory. It is therefore appropriate to review the previous argument on the similarity relationship as an explanation of the law of surface areas. In particular, we can question the generality of the equivalence of the dimensions T and L. Taking up the fundamental work of R. Rosen (in 1983) and that of B. Günther, E. Morgado and R. Jiménez (in 1982, 2003), it is useful to distinguish different types of similarity with a role in physiological allometry, in particular:

– *mechanical similarity*: for the constancy of gravity in our terrestrial conditions $g = LT^{-2} = Cte$, the dimensional equivalence is deduced: $T = L^{1/2}$;

– *biological similarity*: according to Lambert and Teissier's postulate, the equivalence of the two basic quantities is assumed *a priori*: $T = L$.

In addition, there is another very different case when, in addition to strictly metabolic questions, there is the use of mechanical constraints. In such cases, we speak of "*elastic similarity*", the importance of which is known in large organisms, such as tree plants where mechanical constraints link trunk growth in height and diameter (McMahon in 1973, 1983). Using a stress decomposition in two orthogonal directions of length (longitudinal l and transverse d: $l = d^{2/3}$), we must use *Hooke's law* here: $\sigma = E\varepsilon$, giving a linear approximation (units: Pa) of the stress s as a function of Young's modulus of elasticity E and strain ε.

4.2.1.3. *Physiological allometry and growth: Bertalanffy's theory*

The relationships between metabolic allometry and growth were theorized by L. von Bertalanffy in the 1940s and 1960s. This work is part of a general growth model, which posits that the instantaneous rate of weight growth is equivalent to the balance between the processes of anabolism and catabolism, i.e. the building and degradation of body mass y, developing the idea proposed by Pütter in 1920. Each of these elementary processes is defined by a power relationship of metabolic allometry of exponents m and n:

$$\frac{dy}{dt} = a y^m - b y^n \qquad [4.5]$$

Only three types of growth were explained by L. von Bertalanffy, according to the values of m {2/3, 1, 2/3 < 1}, with $n = 1$. This simplifying hypothesis for n, which facilitates the integration of [4.5], implies a proportionality of catabolism with body mass, which would be considered plausible on the basis of observations on weight loss in a diet.

Few developments have been undertaken on this function apart from the cases explained by L. von Bertalanffy, namely:

– $m = 2/3$: mammal and fish growth;

– $m = 1$: larval insect growth;

– $2/3 < m < 1$: growth of gastropod mollusks.

The first case (where there is sigmoid weight growth and linear length growth) is known as the "Beverton and Holt model" in its applications to fisheries management.

Abundant literature continues to be devoted to this issue of physiological allometry in order to agree on a general expression that can account for the relationships (variable according to species) between a particular metabolic indicator and the organism's mass or dimensions. Referring to the many examples published, it is still difficult to conclude, and the debate continues over some of the most frequent values of exponent a, namely: 1, 2/3, 3/4, or a multiple of 1/4. It should be noted that this problem is not reduced to a question of dimensional analysis and similarity, as various other determinants currently under study must be taken into account, such as energy transduction at the level of cell membranes (Demetrius) or the conditions for optimizing circulatory flow in the vascular network in terms of fluid mechanics (West, Brown and Enquist). A review can be found in (Buis 2016). In any case, this theme constitutes a case study highlighting the difficulties in establishing what the status can be of a biological law that is expressed mathematically and supposed to have a certain degree of generality.

4.2.1.4. *Laws of action of a factor: effect/concentration relationships*

Another category of statistical laws concerns the mathematical formalization of the action of a given factor as a function of its concentration. This raises the question in a framework similar to that of chemical kinetics. This analogy is still successful despite the difficulties of biological interpretation resulting from the substantive distinction between molar concentration and "quantity of living matter".

A first case, historically important, concerns the *mineral nutrition* of cultivated plants for the purpose of controlling their fertilization. This question gave rise to various classical expressions, such as the so-called "*law of the minimum*" (Liebig in 1840) and "*law of limiting factors*" (F.F. Blackman in 1905), the practical role of which in agronomy and plant physiology is well known.

The purpose of these laws is to simply state that the yield of a crop is limited by the nutrient that is in the lowest concentration. They were included in the first scientific popularization of agricultural chemistry in the 19th Century, based on the famous image of a barrel whose different staves are of unequal length so that the filling is limited by the shortest stave.

More specifically, under the name of the law of "less than proportional yields", the nonlinearity between dose and effect was highlighted. This followed the old economic considerations of the French minister Turgot (in 1768) on a "law of diminishing returns", linking productivity and the factor of production, a notion that the English economist D. Ricardo sought to theorize (in 1817). Mathematically, this led in 1909 to *Mitscherlich's first law*, which, in chemistry, concerned the action curves of a factor (response vs. concentration).

This is an interesting significant problem with the frequent use of chemical kinetics to interpret a biological relationship. Formally, Mitscherlich's law can correspond to two kinds of chemical kinetics. On the one hand, it is customary to see it as the result of a monomolecular chemical kinetics (hence, its former name of "monomolecular law"), the action of the factor considered being supposed to correspond to the transformation of A into B:

$$A \xrightarrow{k} B$$

As a function of the product formed, $y(t) = [B]_t$, the speed of the reaction is written as:

$$v = \frac{d[B]_t}{dt} = -\frac{d[A]_t}{dt} = k[A]_t = k\{[A]_0 - y(t)\}$$

Analogically, this law was used to model the growth of a variable y as a function of its own value, consisting of linking the instantaneous speed to the growth potential, i.e. to the growth still to be achieved, the latter being assumed to be proportional to the "quantity of remaining factor". This corresponds to cases where growth activity decreases continuously during the process:

$$\frac{dy}{dt} = a(K-y) \qquad [4.6]$$

giving by integration:

$$y = K[1 - b\exp(-at)] \qquad [4.7]$$

For example, in a plant, it may be the tissue growth by cell growth, without multiplication, of a population of meristematic cells, the initial stock of which defines the growth potential.

On the other hand, the same law may correspond to a completely different reaction scheme, that of a balanced reaction:

$$A \underset{k_{-1}}{\overset{k}{\rightleftarrows}} B$$

where the instantaneous state of the transformation is interpreted as a deviation from the equilibrium noted y^*. We draw the expression:

$$y(t) = y^* \{1 - \exp[-(k_1 + k_{-1})t]\} \qquad [4.8]$$

which is formally equivalent to [4.7]. With this second reaction scheme, Mitscherlich's law explains growth as a process of relaxation with respect to the equilibrium value y^*, meaning the cessation of the growth process.

This effect/concentration law obviously does not consider the fact that certain concentrations, above a given threshold, can have a negative toxicity effect when the response function is characterized by the existence of an *optimum*. It was *Mitscherlich's second law* (1928) that allowed this property to be taken into account. Like the previous one, this law is also used as a growth law in complex processes where there is a succession of growth and decline (e.g. in population dynamics).

In this type of problem, another example of a statistical law, *Weber–Fechner's law*, is the relationship between the intensity S of a stimulus and the perception or effect felt I:

$$I = k \ln(S)$$

whose applications in acoustics (audiometry) are well known. It should be noted that this relationship is sometimes referred to as a "pseudo-law" because of its approximation.

4.3. Theoretical laws

Unlike these empirical laws, which essentially result from a statistical smoothing of a set of observations, biology has a few laws called, for convenience, "theoretical laws" (or hypothetical), whose characteristic is to refer either to notions or entities postulated *a priori*, or to explicit hypotheses on the functioning of the process they claim to explain or predict. A first example is *Mendel's laws*, which are well illustrated by their simplicity. These laws initially used a completely imaginary concept, that of gene. On the other hand, they formulate precise hypotheses on the role played by these new "objects" in explaining the distribution properties of characters in a progeny. Two other areas will then be used to present this approach, namely growth laws and population dynamics. It is simply for convenience of language that we will not distinguish between "law" and "model" here.

4.3.1. *Formal genetics*

We have already presented this historical step, which was the elaboration of Mendel's laws (segregation of characters in the progeny of controlled hybridizations) and Hardy–Weinberg's laws (dynamics of a population of genes in panmictic conditions – random mating), which offer us two remarkable examples of theoretical laws. Let us return for a moment to Mendel's laws to highlight their status as theoretical laws.

G. Mendel's hypotheses on monohybridism (only one characteristic at stake) postulated that (i) the first-generation hybrid receives "hereditary elements" made without modification by each of the parents; (ii) two kinds of these determinants correspond to each characteristic, called dominant or recessive, respectively rated A or a; and (iii) any characteristic appears (= phenotype) only if there is the presence of A, or otherwise, it remains latent (dominated). These two categories being of the same number, it follows that the F1 progeny of hybrids of homozygous parents is equivalent to the encounter of the elements, A or a, that they bring during pollination.

Let us quote G. Mendel himself to emphasize his originality: "The differential characteristics of two plants may therefore ultimately be based only on differences in the quality and grouping of the elements"[11]. If the choice of the *Pisum* species and that of the experienced characteristics (which were associated with different chromosomes) were eminently favorable to the expression of this law, as we have previously noted, the fundamental and innovative point is this idea of a combinatorial approach between completely hypothetical elements. In the words of F. Jacob: "The symbolic interpretation of the results becomes the place of articulation between theory and experience"[12].

G. Mendel's symbolism was not exactly that of our current writing of allele pairs (AA, Aa or aa) in coherence with the fact of diploidy, because it assumed that the postulated elements, if they were identical, fused after fertilization. Today, we have the representation of the table (known as the "gamete chessboard") whose column × row entries show the alleles carried by each parent's gametes. The meeting of these elements (self-fertilization of F1 individuals) corresponds in the case of monohybridism to the mathematical development of the product *(A + a)(A + a)*, fixing the distribution in F2, i.e. {1*AA*, 1*Aa*, 1*aA*, 1*aa*}, or the phenotype {3, 1}.

It is known that the existence of exceptions to these theoretical rules by the recombination of characteristics (following a *crossing over* between chromosomes matched during meiosis) does not invalidate the principle of these laws based on the combinatorics of these "genetic elements". It is this same combinatorial method that makes it possible to calculate the recombination percentage and verify its validity.

4.3.2. *Growth laws*

The starting point is the formalization of a growth with only one variable $y(t)$, y being, for example, an organism's size or the size of a population. In these relatively simple cases, the shape of the growth curves is very varied, indicating the existence of a large number of growth laws. In fact, in such a catalog, many formulations use a small number of basic laws, variously arranged with the (somewhat empirical) addition of parameters to improve data adequacy. Some methodological points of this problem (which we have discussed in detail elsewhere) will be included in Chapter 5.

---

[11] In Mayr, E. (1982). *The Growth of Biological Thought: Diversity, Evolution and Inheritance.* Harvard University Press, Cambridge.
[12] Jacob, F. (1970). *La Logique du vivant.* Gallimard, Paris, 224.

In the multivariate case where *y(t)* is vectorial, we have to process a system of growth equations where the intrinsic growth of each variable occurs $y_j(t)$; $j = 1,..., p$ and their interactions $f(y_j, y_k)$ which can be very diverse in nature. The dynamics of biological associations is the classic domain of such models. A wide variety of concrete situations have thus been formalized, depending on whether they involve interactions such as competition, predation (or parasitism) or mutualism, a theme previously discussed with the pioneering work of A.J. Lotka and V. Volterra.

4.3.3. *Population dynamics*

An emblematic case is the association between a prey species and a predator species, an archetype of population dynamics models. Its principle is to write the growth rate of each species as a balance between a so-called intrinsic rate ("as if it were alone") and the intensity of their interaction. This is of sign + for the predator and – for the prey. The predator that depends exclusively on this prey can only decrease in its absence.

From a biological point of view, the most appropriate approach is to write the mathematical hypotheses with reference to specific velocities $(1/y)(dy/dt)$ which express behavior per unit (rather than absolute speeds):

– each species, considered separately, grows according to an exponential law (constant specific velocity);

– their interaction is proportional to their number of individuals (linearity of their interdependence).

These are the simplest assumptions that can be made about the dynamics of this association. This model therefore follows a kind of economic principle. Of course, more refined models have been developed, either by modifying these assumptions or by considering associations of a larger number of species. We therefore highlight different types of stability of stationary states: the existence of a stable limit cycle (ensuring the maintenance of the oscillation of numbers or biomasses) or, on the contrary, either a fixed point, an exclusion or a multistationarity resulting from the partition of the space of the phases into distinct attraction basins.

5

Mathematical Tools and Concepts in Biology

The purpose of this chapter is to review a series of mathematical methods in their applications to biology, according to a choice that would be as representative as possible of the diversity of biological phenomena encountered and the mathematical tools proposed for their analysis. In doing so, we try to avoid both the pitfall of a long technical presentation of each of the tools presented and a list of methods claiming to be exhaustive, but which would simply be a superficial sprinkling of the subject. The essential aim is to show, as well as possible, through a diversity of approaches that sometimes compete, how mathematics can reveal their usefulness when their aim is in connection with this or that biological phenomenon is correctly established. In other words, it is a question of seeing what each method brings and what it implies as basic hypotheses, as a mirror of what each biological process has to offer and what it imposes as experimental constraints on measurement or variability.

As a preamble, it is necessary to return to a general question that sets forth the goal of any mathematization of reality, namely the **duality of representation**: *whether to use mathematics to describe or explain?*

We know that in practice, this dilemma leads to positions that often remain foreign to each other, this irreducibility having nothing to do with the situation offered by the fact of a diversity of explanatory models for the same phenomenon. It is in the field of morphology, when it is interested in the geometric description of biological forms, that such examples are most particularly found. Let us choose for this the famous case of phyllotaxis or study of the spatial arrangement of the sequence of leaves and flowers along a plant axis.

It is indeed interesting to summarize this question of morphogenesis, because due to the diversity of the works that have been devoted to it for a long time, it offers us an illustration of this problem of representation.

5.1. An old biomathematical subject: describing and/or explaining phyllotaxis

Some see phyllotaxis primarily as an architectural problem, while others approach it in terms of processes, seeking to know how this organization or assembly of elements appears and then develops during ontogeny. Of course, the architect is also interested in the forces underlying the building's construction, as D'Arcy Thompson saw it. However, the point of view remains more morphological than physiological, whereas the study of a functioning necessarily integrates time, therefore the dynamics and stability of forms.

From the simple observation of a stem or inflorescence, it is often easy to visualize the existence of fictitious spirals called "parastic", each of which corresponds to a well-ordered series of contiguous elements. Thus, on a sunflower head (inflorescence), 34 sinistral spirals and 55 dexter spirals can be easily distinguished, while on a pine cone, 8 and 13 can be observed, respectively. However, in these examples, these numbers are two successive terms of the Fibonacci sequence. Another observation is that the angle of divergence of two successive units is often close to 137.5°, which is the value of the golden angle $2\pi/\Phi$[1]. Therefore, the question arose to deduce from it a kind of law that would regulate the phyllotactic characteristics by relating them to the Fibonacci sequence. This idea's interpretation, which goes back to A. Braun (1831), continues to be debated, as well as more generally, the meaning that the golden number would have in different dimensional relationships (e.g. the golden number is found in certain geometric characteristics of the DNA molecule, as explained in Figure 3.3).

There is no question of reviewing here the various works on phyllotaxis, a subject in which many authors have been interested since Aristotle and Pliny the Elder, followed by L. da Vinci, before the first precise measurements were undertaken by the Swiss naturalist C. Bonnet (1754). Bonnet underlined the spiral character of the location of the leaves on a stem, from which the parastic notion was deduced as the basic figure of a phyllotactic organization. The subject went beyond botany to crystallography, with the Bravais brothers (1838), who, in mathematicians seeking to link chemistry and geometry, made a representation of phyllotaxis by

---

[1] A statistical study conducted on a thousand sunflower head readings concluded that the 137°.508 value for divergence was well-suited (Okabe, T. (2015). Extraordinary accuracy in floret position of *Helianthus annuus*. *Acta Soc., Bot. Pol.* 84, 79–85).

means of a network of points (lattices) on a cylindrical surface. In addition, from the 1990s onwards, biophysics became interested in the subject (particularly with P. Green) based on the shape of the meristem and the role of mechanical stresses within the cellular foundations of the latter. The interest of A. Turing's reaction–diffusion equations was also highlighted to account for the formation of "patterns" on the caulinary meristem that could determine the emergence sites of the sequence of foliar primordia.

The first explanation proposed for the foliar arrangement along a stem is that of Hofmeister in 1868. This one approaches the problem from a purely spatial point of view, considering that any new primordium is formed in "the largest available space". This idea of space occupation was taken up by various authors, such as the Snows, who, in the 1930s, carried out numerous microsurgery experiments on the caulinary apex. Subsequently, it was decided to add, from a teleonomic point of view, the role of the environment. Phyllotaxis would then be considered by analogy as a control variable contributing to the optimization of the perception of 3D light radiation. However, this is to be qualified, because this perception largely depends on internodal growth, as well as the twisting or nutation of the stem, so that the effect of shading within the photosynthetic foliar system is much later than the generation of primordia. Despite their interest in the "economy" of foliar physiology, let us leave this work to return to the strictly geometric aspects, specifically those that had so interested the first researchers studying the parastic notion.

C. Schimper and A. Braun in 1830 and 1835 set out a hypothesis, which was sometimes considered fundamental, consisting of the assumption that all the elements of a stem are arranged along the same line called the "generating spiral". This qualifier of "generator" implies that there would be a link between kinship (in the sense of the order of succession) and geometry. In other words, it would be confusing a chronological spiral with an ontogenetic spiral. In any case, we can see the difference with the previous idea of several parastics. Botanist L. Plantefol strongly opposed this idea of a single spiral that observation could not verify. In particular, he stressed the need to consider the exact anatomy of the leaf outline, which cannot be reduced to a point, because the underlying "foliar segment" must also be considered. Thus, the foliar dot on the Bravais lattice is not real for a botanist. In any case, L. Plantefol proposed another theory that postulates the existence of multiple leaf helixes, which can better reflect the observed arrangements. The essential question became that of being able to relate the spatio-temporal origin of each of these foliar helixes to a particularity, observable or hypothetical, manifesting itself during the ontogeny of the meristem. For our purpose, it is a question here of underlining the epistemological position of L. Plantefol. While he rightly states the principle of referring to the functioning of the vegetative point (apical meristem), he considers it unnecessary to focus on the search for an "ideal" geometric figure. According to him, "Explaining the phyllotaxis of a stem does not mean finding numerical relationships between the points representative of the leaves and whose position on a stem will ideally have been theoretically rectified to a geometric shape. It means recognizing the relationships that exist on this real stem between the real elements". In other words, it means posing that there is a contradiction between these two objectives, thus advocating *a priori* the irreducibility between the description of the geometrician and the explanation of the botanist.

Although this question of the relationship between physiology and geometry is still far from being resolved, we currently have special insights that renew its approach. Very different points of view from each other, it is worth considering them despite their still theoretical nature.

First, let us summarize what simulations based on the role of dynamic interactions within a set of sequentially generated elements. Two types of simulations were carried out, either experimentally using a physically-controlled device or digitally. In both cases, an ordered set of elements is generated to reproduce apical organogenesis (Douady and Couder 1992, 1996). The experimental device used consists of a circular disk subjected to a vertical magnetic field, and such that the minimum is at the center and the maximum is at the periphery. This disk supports a dish with an oil layer, in the center of which a series of droplets of a ferrofluid are dropped, according to a given period T. Each new droplet or particle, being subject to the repulsion from the previous ones, is placed in a position where this repulsion is minimal. It can be observed that this position is a function of a parameter without dimension $G = VT/R_0$, where V is the velocity of displacement (advection), T the generation period and $R0$ the radius where the particle is placed (the viscosity of the oil is not taken into account). Biologically, T corresponds to the plastochronous (organogenesis rate)[2] and V to the rate of growth of the apical dome after each primordium differentiation. For a high value of T, each new particle interacts only with the previous one, resulting in a 180° divergence corresponding to the phyllotactic facies of alternate opposing leaves. However, we observe a threshold below which its position will depend on the two previous ones, either randomly on one side or the other in relation to the segment connecting them. This results in the direction of rotation of the spiral thus formed. We can describe it as "parastic", but not "generating", in order not to give it an explanatory connotation related to the mechanism of meristematic functioning. A decrease in T results in an increase in the number of particles interacting with any new ones, and correlatively in the number of these parastic spirals.

---

2 The G parameter of Douady and Couder corresponds to the plastochronic ratio P of F.J. Richards (1951), which is the ratio of radial distances to the center of two successive primordia. Since apical growth is exponential and although the temporal and dimensional scales are different, we have a relationship of the type $P = \exp(G)$.

In addition, a sufficiently long numerical simulation (up to 400 iterations to reach a stable regime) uses this principle of antagonism and advection to specify the effect of this G parameter down to very low values. For this purpose, a sequence of material points is sequentially arranged (at a given frequency) on a circle of radius, R, each being subjected to a radial movement at velocity, V. The position of each point is determined according to the overall repulsion or sum of the repulsion potentials (which can be varied) emanating from the different points already generated. The interest of this numerical simulation is to explore in more detail different possible arrangements and to draw a diagram between G and the divergence angle φ. Such a diagram makes it very clear that with the decrease in G, there is a series of bifurcations corresponding to the emergence of parastics, i.e. the existence of thresholds of G conditioning the formation of parastics (so-called "phyllotactic transitions" stages during development). For example, at G values 0.01, 13 and 21, parastics are formed. It can be seen that, below this threshold, there is instability in the angle of divergence φ. Indeed, when G decreases, φ also decreases, but oscillating until it reaches a limit value. This is equal to the golden angle of Fibonacci: 137.5°. Ultimately, the combined existence of multiple repulsive interactions (i.e. a simple and universal "mechanism") between primordia is sufficient to reproduce the two basic phyllotactic characteristics, the number of parastics and the angle of divergence, and to link them to the Fibonacci sequence.

If these simulations remain phenomenological without biological validation, they allow us to give more credit to the connection between phyllotaxis and Fibonacci sequence. On the other hand, it is important to underline the interesting biological connotation of the G parameter. Its variations, as used in physical or digital simulations, are not purely theoretical. Indeed, they are closely related to the ontogenetic evolution of the apex, particularly the non-stationarity of the plastochrone and the modification of the geometric shape of the apical dome. What is quite remarkable is that these ontogenetic variables are presented as coordinates, since they act via this single G parameter.

Let us now complete this question with the contribution of two other points of view that provide, if not direct validation, at least interesting insights, each based on the hypothesis of a generating mechanism.

An original first contribution is offered, in an indirect way, by the development of a 2D cell population formalized by an L-system. The genealogical tracking of the cells whose multiplication is governed by this automaton shows the spiral arrangement of the same family of cells. Without detailing the rules used to determine the spatial and temporal organization of cell divisions (e.g. length of cell cycles, asymmetry and orientation of mitoses), let us say that, from the cybernetic point of view of these formal grammars with ancestral memory, we have here a system generating multiple helices. By simple analogy, it is possible to compare it to what the action of the hypothetical "organizing center" postulated by L. Plantefol would be for its multiple leaf helices. However rudimentary it may be in relation to the phyllotactic organizations observed, these spiral facies are an image that does not seem to have an equivalent.

Another point of view would be that of reaction–diffusion systems, implemented, in particular, by H. Meinhardt in 1975–1980. Already envisaged by A. Turing himself and discussed at the time with botanist C. Wardlaw, this had to go beyond the role of an inhibitor fixing the position of any new primordium. This is obviously in line with, but only partially, all the work inspired by Hofmeister's hypothesis, including the simulations treated analogously by A. Douady and Y. Couder. The question of the Turing activator–inhibitor pair remains. It must be recognized indeed that, following H. Meinhardt's presentation of graphic simulations, the few studies that were devoted to phyllotaxis remain too fragmentary to constitute a basis for a real explanation.

To conclude this review, let us add that recent work[3], broadening the ecological point of view previously noted (shading), is oriented towards the relationship between phyllotactic organization and the physiological optimization of the foliar apparatus. It is indeed necessary to distinguish the phyllotactic disposition of the primordia and the subsequent arrangement of the leaves, because this is more or less modified by any 3D stem growth movement (internodal elongation + nutation). It appears that phyllotaxis contributes to the minimization of the energy required by the torsion of the stem. In addition, the 137.5° divergence would be an adaptive value (fitness) minimizing the number of phyllotactic transitions (bifurcations) that occur during the plant's ontogenesis. Without being an explanation in terms of physiological mechanism, this view opens the field of phyllotaxis by associating it with growth as both contributing to the morphological optimality of ontogenesis. During this phase, the spiral geometry concerns both the stage of organogenesis at the level of the meristem generator and the subsequent stem growth stage, which depends on it. These various considerations clearly show the relevance of a biomathematical approach to such a phenomenon, where morphogenesis and physiology, i.e. geometry and functioning, are closely linked.

5.2. The notion of invariant and its substrate: time and space

Let us return to the notion of the **invariant**, the importance of which we emphasized in Chapter 2, but which presents itself very differently according to the discipline.

---

3 Okabe, T. (2015). Biophysical optimality of the golden angle in phyllotaxis. *Sci. Reports*, 5 (art. 15358).

In mathematics, the notion of invariant is linked to that of transformation. An invariant is a characteristic that does not change as a result of a given transformation (such as a change in coordinates). Let us give two simple examples. The angles and distance ratios are invariant with the transformation rotation, translation and reflection. The Euclidean distance is invariant with orthogonal rotation. Another well-known case is that of all conics (sections of a cone by a plane according to its position). These curves, defined in Cartesian coordinates $\mathbb{R}^2$ by a general equation of the second degree, have as invariant the quadratic expression $ax^2 + bxy + cy^2 = 0$, which constitutes its signature. Indeed, depending on the value of the parameters, it is the value of the discriminant $\Delta = b^2 - 4ac$ (<0, 0, >0) which characterizes the type of conic, respectively, an ellipse, a parabola or a hyperbole. The interest of the mathematical invariant is thus to establish a relationship between different objects, to make them a kind of typology.

Biology, while it speaks rather little explicitly of invariant, offers us some remarkable examples. In general morphology, we have the notion of an organizational plan. For higher organisms, this was one of the main themes of debate in the 19th Century, with F. Cuvier, É. Geoffroy Saint-Hilaire and E. Haeckel. In biochemical and physiological terms, we have very fundamental notions such as the universality of basic metabolic cycles, cellular energetics (role of ATP), macromolecules (DNA and RNA) or genetic code (with very rare exceptions). Following cell theory (the premises of which date back to the 17th Century with A. Van Leeuwenhoek and R. Hooke), this notion of invariant constitutes the very condition of what is called the "unity of the living". We remember the famous statement (attributed to J. Monod) that "what is true of the colon bacillus must be true for the elephant". This aphorism, concerning the genetic code, is repeated moreover on the thinking of biochemists A. Kluyver and J.L. Denker who, in 1926, said that "from elephants to butyric bacteria, everything is the same". In a completely different way, we will try to clarify how the general idea of invariant can be expressed in biology through some remarkable mathematical properties.

In biology, it is necessary to associate with this concept of invariant what constitutes its substrate, i.e. the notions of **time** and **space** as biology uses them. They will serve as a guideline for the development of connections between biology and mathematics, which we will then have to specify on the basis of a few illustrative examples showing the mathematical methodology at the service of biology – both on the technical level of the tools and in the manner of asking the questions.

5.2.1. *Physical time/biological time*

To speak here of time, without any epistemological or metaphysical consideration, is not an empty question. This old and recurring problem deserves a word about it in relation to our subject. The question arises as to the choice of the temporal reference frame: do we simply adopt physical time or sidereal time, as we do for the study of a Galilean reference frame movement (uniform and isotropic space, uniform time) or the advancement of chemical kinetics? On the other hand, can we define what would be a biological time, specific to the phenomenon studied and what would be its value? In other words, can we say: "To each their own time?"[4]

On this subject, we can mention the following three approaches because of their merit in not limiting themselves to mere speculation, but seeking validation of the proposed model.

5.2.1.1. *The physiological time of Lecomte du Noüy (1936)*[5]

With a highly practical aim concerning the healing of war wounds, P. Lecomte du Noüy sought to formulate the duration of wound healing by considering that two distinct variables – both of a temporal nature – should be taken into account: (i) physical time as a unit of measurement to quantitatively express the kinetics of wound coverage and (ii) the condition of the wound, and particularly its age, which determines its own capacity to react. In other words, there would be two temporal components of physical time: (i) its current course at the level of the injury itself and (ii) the temporal mark that remains inscribed in the injured person as an integration of the elapsed physical time as a descriptor of his physiological state. This last component is not related to the instantaneity of physical time, but to a duration that has just occurred in the organism studied, according to the philosopher H. Bergson's well-known idea.

5.2.1.2. *Backman's biological time (1942)*[6]

The objective here is to find a growth function that would be invariant in order to apply to different species. On the one hand, it would be independent of adult dimensions (standardization). On the other hand, its kinetics would differ, depending on the species, from a transformation of the measured variable, allowing us to highlight the singularities of development. This would be tantamount to obtaining a kind of intrinsic phenology of fundamental stages. This question arose in the 1920s

---

4 We use here the title of the book by Pacault, A., Vidal, C. (1975). *À chacun son temps*. Flammarion, Paris, which exposes time in its various manifestations, from physics to psychology and economics.
5 Lecomte du Noüy, P. (1936). *Le Temps et la vie*. Gallimard, Paris.
6 Backman, G. (1942). Das Wachstum der Bäume. Das Wachstum einiger Kulturpflanzen. *Wilhelm Roux' Archiv*. 141, 455–499, 770–816.

in Brody, who was one of the pioneers of biological growth laws – particularly with regard to the weight growth of animal species. It was developed by G. Backman, who clarified this issue of major ontogenetic stages as invariants. This crucial question was occasionally taken up by some authors, without yet reaching an experimental validation, coming up against the difficulty of a good temporal signage of these singularities.

5.2.1.3. *Plastochronic time in higher plants (according to Erickson 1957)*[7]

The development of a higher plant can be described by taking as a marker the occurrence of the formation of a new morphological unit (generation of a new module {node, leaf, axillary bud}). It should be recalled that "plastochronous" refers to the duration (in physical unit) of generation of a new leaf primordium, i.e. the duration of the meristematic functioning cycle of the caulinary apex. In fact, to avoid observation by dissecting the bud, non-destructive measurements such as the acquisition by the young leaf of a reference size perceptible macroscopically are agreed. Another difficulty is that this principle is based on a stationary regime of meristematic functioning, which is generally not the case, the plastochrone varying more or less during ontogenesis. In any case, botanical literature still refers to this type of intrinsic time.

5.2.1.4. *The change of state of the system*

More generally, but still very theoretically, this idea of a time that would be specific to the object under study can be referred to, according on R. Vallée[8], in the occurrence of a change in the state of the system. It means considering that physical time or duration is only data external to the process. What is essential is the perception of the system itself. According to this point of view, the passage of time must be related to the change in the state of the system, i.e. the definition of an intrinsic duration between two measuring moments t_1 and t_2: $d(t_1, t_2) = \int_{t_1}^{t_2} \left\| \frac{dy}{dt} \right\|^2 dt$

This implies having an adequate formalization of the velocity of the process.

5.2.2. *Metric space/non-metric space*

Due to the generally inhomogeneous nature of biological objects, the notion of "field" is in principle of particular importance for the study of spatialized processes. Certainly, we do not have in biology what would be the equivalent of what physics knows with electromagnetic fields and the formalism of J.C. Maxwell's equations.

---

7 Erickson, R.O., Michelini, F.J. (1957). The plastochron index. *Amer. J. Bot.* 44, 297–304.
8 Vallée, R. (2005). Time and systems. *Kybernetes*, 34, 1563–1569. See Buis, R. (2016). *Biomathématiques de la croissance*. EDP Sciences, Les Ulis (web companion, Chapter F).

On the other hand, growth and morphogenesis are quite dependent on field analysis. For example, in plants, growth is almost always very unevenly distributed within the different organs. Whether in dimension 1 (root elongation), 2 (extension of a planar leaf blade) or 3 (functioning of the meristem of a bud)[9], there are clear growth gradients (uneven distribution of local activity) and anisotropy (variation in activity depending on direction). Animal embryology leads us to similar considerations, to which is added the role of cellular migration. All growth and morphogenesis are by nature spatio-temporal processes that take place in a space considered "Euclidean". The study of biological forms also uses a Euclidean metric, as we have seen for dimensional allometry establishing growth disharmonies between different regions of a tissue or organism. However, the question is complicated by the theory of shape transformation as illustrated by D'Arcy Thompson using curvilinear coordinate systems (see Figure 3.5). Let us also add the use of modern morphometric methods (mathematical morphology, image analysis) in biology, where specific metric questions are raised.

5.2.2.1. Metric space

According to a completely different objective, the biologist may be required to work in a space with p dimensions, say $\mathbb{R}^p$, for example, for the representation of a statistical correlation diagram of a series of individuals on which p variables or characteristics have been measured. The Pythagorean theorem then remains valid in a generalized form. The distance between two individual vectors x_i and $x_{i'}$ or norm of their difference is written (the operator being the usual scalar product I):

$$d^2(x_i, x_{i'}) = \|x_i - x_{i'}\|^2 = \|x_i - x_{i'}\|^T I \|x_i - x_{i'}\| = \sum_{j=1}^{p}(x_{ij} - x_{i'j})^2$$

However, by staying within these kinds of questions, there are other metrics, such as those used in the analysis of multi-dimensional data. First, let us note the Euclidean metric used in the *principal component analysis* of H. Hotelling (1936), an archetype of *factor analysis*. This gives the same "weight" to the n individuals, in the form of a diagonal matrix, whose elements are all equal to $1/n$, used for their spatial representation of the principal components (calculation of the inertia of the cloud of individuals in relation to the different main axes). The situation is different with *correspondence factor analysis*, which works differently because instead of raw data, it uses a contingency table, consisting of a set of line and column profiles (resulting from weighting according to marginal distributions). The particularity of this method (which makes it successful) is to allow a joint representation of

---

9 Note in passing the fractional dimension of natural fractal objects such as highly branched systems (bronchial arborization, vascular network, root system, mycelial filamentous system).

individuals and variables in the same vector space. This implies taking into account the differences in the marginal numbers of rows and columns. In order to avoid giving more importance to some of them, a special metric, called "metric of the χ^2" (chi-2). The distance between rows i and k (on q columns) will therefore be:

$$d_{\chi^2}(i,k) = \sum_{j=1}^{q} \frac{n}{n_{.j}} \left(\frac{n_{ij}}{n_{i.}} - \frac{n_{kj}}{n_{k.}} \right)^2$$

notations of the type $n_{.j}$ designating marginal distributions.

Another case is that of *canonical analysis*, a generic term for the analysis of several series of data. H. Hotelling was the first to investigate the relationships between two data sets, i.e. two sets of individuals on which p variables were measured. One of its well-known applications is *discriminant factor analysis*, some of whose algorithms allow the choice of m variables, $m \leq p$, maximizing discrimination (= distance) between the two series. This optimization is achieved by a linear combination of the characteristics considered. For educational purposes, we can cite a simple example of an application made by R.A. Fisher in 1936, concerning the discrimination of three species of *Iris* on the basis of $p = 4$ flower size characters (data published online). In such studies, a non-Euclidean metric called "Mahalanobis distance" is used as a measure of the *similarity of two data sets*. Instead of giving equal weight to different individuals, a weight is given according to their inverse dispersion. This is the same idea as judging data that are far removed from the majority of the sample as "aberrant" or "atypical". The weight of data with a low probability of occurrence is therefore minimized. In the case of a cloud of observations measured on interdependent variables, the distance between two vectors x and y is defined by taking into account the dispersion of the variables and their correlations, namely:

$$D_M(x,y) = \left[(x-y)^T \Sigma^{-1} (x-y) \right]^{1/2}$$

where Σ is the variance–covariance matrix, the exponent T designating the matrix transposition.

If $\Sigma = I$ (identity matrix), we find the Euclidean distance mentioned above.

It should also be noted that the multi-dimensional data analysis involves various techniques using non-Euclidean distances to measure similarities. An example is the dynamic cloud method of E. Diday, consisting of obtaining step-by-step an optimal aggregation of groups of individuals from nuclei that can be reworked, or numerical taxonomy algorithms that use particular metrics that are unnecessary to detail here (see Sokal and Sneath 1972). Apart from these phylogenetic considerations, we can mention its interest in the study of the polymorphism of current natural populations (resulting from gene flows *in situ*), such as the interesting example of the forage grass *Panicum maximum*, with the study of about 30 characteristics to clarify its heterogeneity and especially to deduce similarities in the context of a genetic study of grain domestication[10].

5.2.2.2. Non-metric space

In biology, we now know that everything related to space is not reduced to these classic and relatively simple questions of metrics. Thus, molecular genetics distinguishes between different kinds of "distances", depending on the type of *genetic mapping* considered (whose name is not always precise). Distances can be related to two different levels: that of a given chromosome or that of an entire genome. In the first case, we work on the linear sequence of the locus of the same chromosome. The physical distance between two loci can then be simply estimated by the number of nucleic base pairs (*bp*) separating them. However, another type of measurement is also used, based on the crossing-over phenomenon, to which any chromosome is subject with its counterpart during mitosis (at the meiotic stage). This results in an exchange of locus called "recombination". The probability of segment exchange between paired chromosomes is proportional to its length, i.e. the distance between the two loci that define it. This distance, which is different in nature from the previous one, is measured by a recombination frequency. The unit is centiMorgan: 1 cM = 1% recombination. Finally, at the global level of the genome, we are interested in another kind of distance. It is not about two linked loci, but between two "molecular markers". By marker, of which there are different types, we mean here a detectable DNA sequence whose polymorphism makes it possible to characterize an individual genome. Thus, for the biologist, this term of distance may no longer be an entity of a geometric nature, but may address a variety of physical realities beyond the previous simple question of metrics.

However, the most remarkable thing is the importance that should be given at this time to the *topological aspects* of certain processes. In this context, we can recall the classic question of the different types of protein structure. We thus move from the linear sequence of amino acid linkages to the conformation properties (tertiary and quaternary structures) resulting, on the one hand, from the folding of peptide chains and, on the other hand, from the existence of typically arranged subunits. Then again, with the evolution of the conception of what is meant

---

10 Work by Pernès *et al.*, at ORSTOM in 1960–1980; see Pernès, J. (1983). La génétique de la domestication des céréales. *La Recherche.* 146, 910–919.

by "gene" that we have talked about, we must rethink G. Canguilhem's general purpose about what has become of the knowledge of life.

PRINCIPLE.– Knowledge of life "does not resemble architecture or mechanics, as it was when it was simply anatomy and macroscopic physiology. But it looks like grammar, semantics, and syntax [...]. To understand life, it is necessary to use a non-metric theory of space, *i.e.* a science of order, a topology." [11]

We should not neglect the fact that mechanical constraints are spatially linked to the growth and development of large organisms such as tree species.

5.2.3. *Multi-scale processes*

Let us leave these general considerations behind and highlight this characteristic that biology is now becoming aware of, namely that certain phenomena have the particularity of taking place on **several time and/or space scales**. With respect to time, it has long been known that some oscillatory systems exhibit a characteristic slow dynamic/fast dynamic association, as seen in Van der Pol's well-known system, where there may be a regular succession of these two behaviors (Figure 5.3). Observation of certain biological processes attests to their reality in various situations, such as the rhythmic morphogenesis of the acrasial amoeba *Dictyostelium*, which can change from a free form to cellular aggregation.

Space itself can be subject to this kind of duality, which must distinguish between macro- and microscopic structures. For example, the behavior of a material is related to both (i) its macroscopic aspect seen as a large-scale continuum and (ii) the existence of microstructures at a crystalline microscopic scale where the aging and degradation processes of the material take place. From this rheological point of view, we have an equivalent with some so-called "hierarchical" biological structures, such as bone tissue. More generally, it is the interest of a statistical physics approach that takes place in biology.

5.3. Continuous formalism

Let us consider the representation of a biological process by a dynamic system (ordinary differential equations, partial differential equations or integro-differential equations) that can express some fundamental characteristics. The main focus is on

---

11 Canguilhem, G. (1983). *Études d'histoire et de philosophie des sciences,* 5th edition. Vrin, Paris, 362–364.

(i) the existence of spatial or temporal singularities whose mathematical importance reflects an essential biological property (local extremum of an activity), (ii) the possible dependence on initial conditions in cases of multistationarity and (iii) the role of exogenous variables or fluctuations on the stability of the steady state. The dynamics of the last two cases is conditioned by the existence of a bifurcation or qualitative change in the dynamic regime. Let us describe this methodology in some detail, the biology of which has been progressively used according to the nature of the phenomenon.

5.3.1. *Dynamics of a univariate process*

The process can be approached by a single global variable, which is supposed to be sufficient to describe it as do the basic kinetic models used in the classical formulation of growth laws:

$$\frac{dy}{dt} = f(y,P,t) \qquad [5.1]$$

If the speed of the process depends only on constant **P** parameters, the model is called "autonomous deterministic". We assume here an instantaneity of the cybernetic relationship, the current state *y(t)* determining its own variation in sign (growth or decrease) and value (acceleration or deceleration):

$$y(t) \xrightarrow{k} dy(t)/dt$$

Here are some common examples of the formalization of a growth based on the assumption that the instantaneous velocity *dx/dt = f(x)* is autonomously defined. Some growth laws postulate the limit value (theoretically asymptotic for a simple mathematical constraint) set *a priori* as a parameter (representing, for example, the biotic capacity of the environment, often denoted *K*).

P.F. Verhulst's logistic model is the archetype of this kind of formulation by posing as a velocity equation: $\frac{dx}{dt} = ax\left(1-\frac{x}{K}\right)$, i.e. the ago-antagonistic association of a potentiality qualified as Malthusian (exponential) and a deceleration braking, giving a symmetrical sigmoid curve. Various avatars, called "generalized logistics", have been proposed to generate asymmetric sigmoids with a variable position of the maximum activity time (inflection point), for example: $\frac{dx}{dt} = ax\left(1 - \left[\frac{x}{K}\right]^n\right)$ or $\frac{dx}{dt} = ax^m \left(1 - \frac{x}{K}\right)^n$.

Another kind of formulation poses the growth process as the balance between anabolism and catabolism, both functions of a power of the growing variable: $\frac{dx}{dt} = ax^m - bx^n$ (Bertalanffy, Blumberg), the limit value is not fixed *a priori*.

5.3.2. Structured models

Despite the interest of these univariate models, which were and still are the basis for various models, clearly, they can only lead to a strong approximation of reality, because they do not take into account the heterogeneity of the population studied. Beyond a first study, it is therefore necessary to address so-called structured models, based on a partition (or stratification) of the system or population into different classes or compartments, each of which is relatively homogeneous (we can speak of functional equivalence of individuals of the same class) according to such state criteria. In addition to the specific kinetics of each category, the transition rates between classes must therefore be considered.

5.3.2.1. The univariate function of Gompertz

Let us illustrate this principle with the famous univariate function of Gompertz (competitor of logistics) that can be defined as a non-autonomous model by an exponential decrease in the specific velocity over time:

$$\frac{1}{x}\frac{dx}{dt} = k\exp(-at)$$

This function is of great interest to biologists outside its initial demographic field, particularly because of its applications, which have become classic for malignant growth. It is precisely in relation to this type of application that this function was rewritten by different authors in a form that takes into account the heterogeneity of a total cell population N subject to tumor evolution. For this purpose (we summarize), two categories of cells were considered: the *P*-proliferating cells and the *Q*-quiescent cells, according to Figure 5.1.

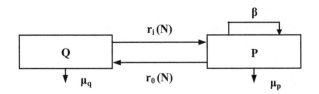

Figure 5.1. *Two-compartment diagram of the Gompertz model (depending on the version of (Kozusko and Bajzer 2003))*

formalized by the dynamic system:

$$\frac{dP}{dt} = [\beta - \mu_P - r_0 N] P + r_i N Q$$
$$\frac{dQ}{dt} = r_0 P - [r_i + \mu_Q] Q$$
$$N = P + Q$$

where μ_P and μ_Q are the mortality rates of these two cell types, β the division rate and the transitions r_0 and r_i between these two compartments depending on $N(t)$. Various simulations of this model were performed depending on the value of the parameters and their validation for different types of malignant growth.

Of course, this principle of discrete structuring into state classes applies to many other growth functions. Thus, the logistic classic is presented as belonging to kinetics of different cellular categories, taking as a stratification criterion the lifetime (or generation time) of each sister cell of an asymmetric mitosis (which is a common characteristic of the ontogenesis of filamentous organisms, fungi or algae). A second condition criterion concerns the consideration of the evolution of these lifetimes during the development of the organism (senescence effect) (Buis and Lück 2006).

5.3.2.2. *Multi-compartment models*

This principle of structured models, which we have just shown in a simple way, is used for many kinetics studies, whenever there is a relevant basis for compartmentalization. One example is the many models developed in pharmacodynamics (human or veterinary biology). This flexibility provided by compartmentalization can be illustrated with the example of chemostat cell growth. The experimental device used (reactor) allows continuous growth monitoring (by sampling in each class, and partial renewal called "dilution") with control of nutrients and culture medium inputs (Figure 5.2).

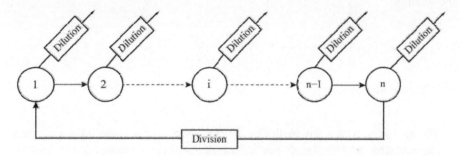

Figure 5.2. *Growth in chemostat of unicellular green algae Cryptomonas sp., compartment model (after J. Arino 2001))*

The dynamics of such a monospecific population is addressed by stratifying into different classes on a given criterion, which is here the stage of growth. In addition to the class transitions by growth, there are also the dilution outputs corresponding to the partial renewal of the culture medium. The growth of the biomass xi of each class is written as:

$$dx_i / dt = \alpha_{i-1} x_{i-1} + \mu x_i + \alpha_i x_i$$

The overall dynamics thus integrates the kinetics of each of these categories and the class transitions, knowing that these result from processes of activation, differentiation, quiescence or cellular mortality.

5.3.3. *Oscillatory dynamics*

The use of differential formalism can lead to a wide variety of dynamic behaviors. Indeed, unlike the previous univariate growth laws, which are asymptotic in nature, some systems can exhibit large oscillations with great qualitative diversity depending on the system studied. Indeed, depending on the case, we observe either a damping of the periodicity or, on the contrary, their increase (explosion by instability), or a chaotic dynamic (irregularity of the sequence of periods, sensitivity to initial conditions, non-predictability) (see Goldbeter 1990; Françoise 2005). Not being able to detail this diversity of behaviors, let us cite an illustrative example particularly interesting in physiology. These are systems that can have two types of dynamics along their trajectories, namely an ordered sequence of "slow" and "fast" dynamics. The Liénard-Van der Pol oscillator is the classic model. It is defined by the system:

$$\frac{dx}{dt} = f(x,y) = \frac{1}{\varepsilon}\left(x - \frac{x^3}{3} + y\right)$$

$$\frac{dy}{dt} = g(x,y) = -\varepsilon x$$

Figure 5.3 illustrates this particular dynamic. We have a stable limit cycle that is characterized not only by its immediate singularities (extremums), but also by pre- and post-extremum variations. Here, it is the parameter ε that determines the type of dynamics, in particular, the property that any maximum can be framed by two very different phases, – at high speed, then at low speed. It should be noted that, depending on the value of the parameter ε, this system can generate a completely different behavior, that of regular sinusoidal oscillations.

This type of dynamic system is of great interest to certain physiological processes, such as the rate of heartbeat (slow ventricular waves/rapid) or the propagation of nerve impulses (action potential). More generally, this duality is found in the combination of metabolic variables (generally fast dynamic) and genetic variables (transcription and translation, slower dynamic).

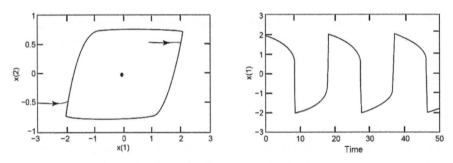

Figure 5.3. *Van der Pol Oscillator. $\varepsilon = 0.1$. Left: stable limit cycle generated from two different initial conditions; right: asymmetrical shape of the oscillations*

This characteristic behavior is quite distinct from the dynamics determined by a delay effect that occurs when the instantaneous velocity depends both on the current state of the variable *y(t)* and on one or more previous states *y(t - τ)*. The delay *τ (lag)* can correspond, for example, to a period of maturation or incubation that must be considered in population dynamics. We know various logistics with delay (e.g. the Cunningham model) whose dynamics can be varied, including oscillatory behaviors.

For this type of problem, it is interesting to add the interest of moving from a univariate form to a more general form of a dynamic two-variable system, the aim being to be able to draw an exhaustive picture of the different possible dynamics, detailing different cases of stationarity. For this, we consider two state variables: (i) the variable measured y itself (a dimension or a number, for example) and (ii) the specific velocity $(1/y)\,(dy/dt)$. The reason for this formalism is to be able to envisage a wide variety of initial conditions. In other words, these two state variables are considered as separate variables at the beginning of the process, giving a potentially important role to the initial state of the object being studied, as indicated by many experimental situations. Let us take the example of a mycelial or algal filament, where we know that growth and morphogenesis of any neoformed cell depends on its state at birth: its size, its relationship with the mother cell it inherits in part, its position within the filamentous system, etc., from which will result not only its adult dimensions, but also its morphogenetic becoming, which may or may not be the seat of budding that will be at the origin of a lateral branching.

In the same vein, let us add that such a value of a process velocity does not have the same biological significance according to whether it is increasing or decreasing, i.e. according to the sign of its acceleration (second derivative). This is taken into account by the logical kinetics approach (see section 5.7).

REMARK.– With regard to the choice of a model, numerous works have been done to compare different mathematical functions for a given process, mainly interested in their respective quality of adjustment to the data. In reality, the issue goes well beyond this statistical point of view of a significant correlation between estimates and observations. Among other examples, we can cite the competition for the formalization of a growth process, the logistics law of P.F. Verhulst with other functions. We know of a few cases where this is a commonplace formal equivalence. For example, the hyperbolic tangent function (which was proposed for growth kinetics) is exactly the same as the quadratic velocity equation of logistics. On the other hand, the comparison with other very different functions is a matter of debate. This is seen, for example, with the proposed use of the Laplace–Gauss normal law of probability (distribution function). In this case, the decision is clear. On this choice, indeed, it is necessary to underline the interest of the logistics law, which is to propose an interpretation of the process (interpretation which can therefore be put to the test). In comparison, its substitution by the normal law of probability remains, in general, purely technical, without interest of the biologist seeking a hypothesis on the possible explanation of the process.

5.3.4. *On the stability of dynamic systems*

5.3.4.1. *Structural stability, an implicit assumption of dynamic systems*

We will remain in relatively simple cases to present now some important remarks on the stability property presented by stationary states. On this subject, the term **structural stability** of a model refers to a major property of a dynamic system that is of the same nature as the robustness of a behavioral or statistical test, i.e. leading to the reliability of the conclusions displayed by the model. This means that there is relative independence from local variations or disturbances affecting parameters or the environment. We will look at the theoretical situation before presenting some critical examples of this notion, adding the role of exogenous variables exercising the role of process control or command, i.e. intervening on the quality of stationary states, a question related to the general question of optimality.

From a mathematical point of view, this term strictly refers to topological considerations. It means that a deformation of a vector field (leading to a change of norm) does not modify its topological characteristics (we speak of "homeomorphism of two vector spaces" in the sense of one-to-one and two-continuous application). Taking up R. Thom's remarks (Thom 1972, p. 31), the term applies to any object or figure that common sense refers to as a "subjectively identifiable form". The reason for the name is that any object is "always subject to disturbing influences from the external environment", influences which, while making some modifications, leave the object in the same equivalence class, i.e. there is permanence of the "form" in the usual sense of the word. R. Thom adds that "the hypothesis of structural stability of a process appears as an implicit postulate of any observation".

Thus defined, this notion would therefore have to be distinguished from the stability in the dynamic sense that we typically use, and which simply refers to the well-marked properties of a steady state (stable versus unstable, plus cases of the so-called "neutral" or "conditional" stability). In accordance with well-established practice, we will accept this extension, focusing on the quality that such a dynamic system can have (its formal structure and the domain of the numerical values of its parameters) as a means of highlighting our essential objective, which is to achieve the *invariance of the intrinsic characteristics of the biological process studied.*

Let us illustrate this stability issue using a simple two-species predation system (system resource = prey/consumption = predator). We know that the archetypal model of such a biological association of prey–predator or host–parasite is that of Lotka–Volterra, already mentioned many times because of the historical reference it continues to have in epistemology. It should be recalled that the original purpose was to seek an explanation of D'Ancona's observations in the Adriatic Sea on the increase in the density of certain predatory fish species (selachians) following World War I, which amounted to protecting prey by reducing fishing.

This very simple model, which generates self-sustaining oscillations, has the disadvantage of its strong dependence on initial conditions, which determine the period and amplitude of oscillations. It cannot, therefore, identify an autonomous dynamic that is characteristic of the interactions between the two species.

Let us recall its formulation on the basis of the following assumptions: (i) the potential demographics of each species (prey x and predator y) considered in isolation follow an exponential law (prey growth and predator decline); (ii) the predation rate ("functional response" or prey consumption per unit of predator) is proportional to the prey density; (iii) the predator growth rate ("numerical response") is linear with respect to the prey density. This gives the following velocity equations:

$$\frac{dx}{dt} = ax - bxy$$
$$\frac{dy}{dt} = -cy + dxy \; ; \; a,b,c,d > 0$$

This very simple system has a single stationary state S, intersection of the two isoclines parallel to the axes:

$$\text{isocline } x : dx/dt = 0 \Rightarrow x^* = c/d$$
$$\text{isocline } y : dy/dt = 0 \Rightarrow y^* = a/b$$

S is not an attractor r (neutral stability): any position of the current point in the vicinity of S is normally maintained in an orbit close to S. On the other hand, any minimal disturbance (displacement outside a given trajectory determined by the initial conditions) leads to a change of orbit without returning to the previous trajectory.

These disadvantages have been addressed by modifying (somewhat empirically) the above basic assumptions. In particular, the choice of more realistic assumptions focuses on the potential growth of each species, which can be described as asymptotic in nature such as logistics (rather than Malthusian). On the other hand, based on various observations depending on the species and rejecting the linear approximation accepted above, the predation rate is assumed to be limited, for example, with a negative hyperbolic or exponential saturation effect. Practicing a kind of experimentation on the initial formalism of Lotka–Volterra, the literature proposes different alternative models (May, Tanner, etc.) for which a mathematical stability study should be carried out.

Kolmogorov's theorem precisely defines the necessary stability conditions for a predation system according to the following general form:

$$\frac{dx}{dt} = x f(x,y)$$
$$\frac{dy}{dt} = y g(x,y)$$

[5.2]

– In summary, let us say that the existence of either a stable stationary point or a stable limit cycle is determined by the following main conditions:

– f and g are continuous functions with continuous first derivatives (at least on the positive domain of x and y);

– $\partial f / \partial y < 0$; $x(\partial f / \partial x) + y(\partial f / \partial y) < 0$;

– $\partial g / \partial y < 0$; $x(\partial g / \partial x) + y(\partial g / \partial y) > 0$;

– $f(0,0) > 0$.

5.3.4.2. *The paradox of enrichment: the Rosenzweig–MacArthur model*

One of the well-known forms of the system [5.2] is the Rosenzweig–MacArthur model, which was at the origin of what is called the "paradox of enrichment". We know that predation can be considered as a factor regulating natural ecosystems. We observe this with the existence of oscillations affecting both prey and predator, a qualitatively well-documented behavior in different associations of very diverse species. The effect of such variations (which some models can generate through a stable limit cycle) allows the maintenance of a dynamic equilibrium ensuring the permanence of each species without the risk of exclusion of the prey.

The reality is not so simple, however, because such systems, due to their structural instability, may be subject to *destabilization*. This is the well-known case of eutrophication of an environment: an increase in nutrient resources does not only have the primary effect of stimulating prey growth, but can also destabilize the system itself. This amounts to considering that this enrichment has the nature of a disturbance that the ecosystem cannot regulate, without the possibility of evolution towards a new situation of equilibrium between species. This is because nutrient enrichment (natural or induced) is equivalent, for the model that is supposed to explain the dynamics, to a change in the value of certain parameters of the growth of the prey species, and therefore interactions related to the rate of predation. As a result, as some models may specify, the combined effect of these two components is sufficient to modify more or less strongly the value or even the occurrence of the equilibrium state and especially its possible stability. In other words, instability must be seen, not as an exception or a paradox, but simply as an intrinsic property of the model.

Let us schematically look at the elementary association one prey–one predator, a simple case where we can easily specify how the relationship between the prey's growth function and its consumption function is presented. Rather than the general form given in [5.2], use the following variant of the Rosenzweig–MacArthur model with logistic growth of the prey and hyperbolic limitation of the predation rate:

$$\frac{dx}{dt} = x\left(1 - \frac{x}{c}\right) - \frac{xy}{1+x}$$
$$\frac{dy}{dt} = by\left(-a + \frac{x}{1+x}\right)$$
[5.3]

Let us graphically illustrate this dynamic using the plot of the corresponding isoclines whose intersection S is governed by the numerical value of the parameters[12].

Figure 5.4 simply indicates the existence of different cases of stability of the theoretical stationary point S: no equilibrium (A), a stable equilibrium (B) and (C) and an unstable equilibrium (D). By moving the vertical isocline $dy/dt = 0$ from right to left, the dynamic regime change is visualized as it passes from the top of the parabolic isocline $dx/dt = 0$: in the region to its left, point S becomes unstable, changing the type of demographic evolution of each species.

The corresponding phase portraits would show the details of the trajectories. In case (B), the trajectories converge towards the stationary point S attractor. In (D), where S is a repellent (of the unstable focus type), the trajectories evolve towards a stable limit cycle (permanent oscillations of the two species). Case (C) is still of the point attractor type, but due to the structural instability of the system, the occurrence of a disruption may lead to a switch to the case (D).

---

12 For example, you can consult the CNRS *Experimentarium Digitale* interactive site (digital examples).

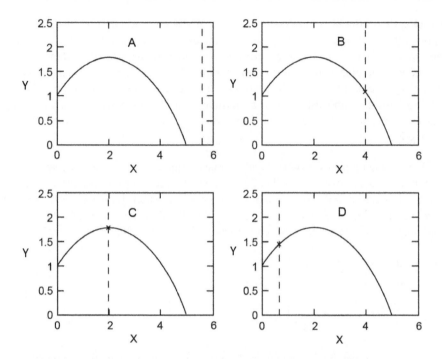

Figure 5.4. *Rosenzweig–MacArthur predation model: b = 1; c = 5; A: a = 0.85; B: a = 0.8; C: a = 0.6; D: a = 0.4; *: equilibrium point S*

Of course, before these remarks, we must question the relevance of the model itself, reminding us how empirical the attempts to improve the initial Lotka–Volterra model were. For example, abandoning the strong (or even unrealistic) assumption of Malthusian growth for the prey species is not necessarily self-evident, because predation limitation may justify neglecting its autonomous self-limitation when it is very low in relative values compared to the intensity of its consumption by the predator (moreover, this type of approximation is often accepted without difficulty in many models). However, above all, the logistics of P.F. Verhulst, which is often used, is not the only appropriate formalism. Similarly, it is empirically known that the predation rate can be formulated very differently (there is no general law linking consumption and abundance). These remarks underline the importance of the *a priori* choice of the components of the model on which the type of dynamics will depend. The models are very often built on the basis of components that appear "plausible" in the absence of an exact mathematical description of a particular process. However, it appears that some models are very sensitive to even minor changes in formalism. We talk about *"super-sensitivity to structure"*.

All this can lead to uncertainty about the quality of the model, and therefore about the possibility of a good prediction. Various studies periodically[13] highlight this, both by mathematicians and biologists, motivated by a desire for formal rigor or simply a requirement for calculation. This has been denounced many times, especially concerning the modeling of biological associations. In this field, the use of functions that are very similar and have the same capacity to describe phenomenologically a nutrient consumption process can lead to quite different behaviors at the level of the integrated system (through interactions).

Nevertheless, the mathematical formalization of the problem is the only way to be able to detect the conditions of existence of a relational invariant in the phenomenology of the studied phenomenon. What is now practiced with an experimentation called *in silico*, completing in its own way what was the approach of *in vitro* experimentation compared to observation *in situ*.

5.3.4.3. Stability and optimal process control

A question related to the dynamic quality of stationary states and the principle of optimality concerns the role of control variables. The problem of stability then arises to look for the control surface whose morphology may include sudden breaks or jumps in the value of stable stationary states.

A remarkable formalization is given by R. Thom's theory of catastrophes, which highlighted the existence of seven cases called "elementary disasters". The simplest is what is called the "cusp" corresponding to the system: a variable u and a control parameter a according to the equation $F(u; a) = u^3 + au$. A slightly more complicated case is called the "dovetail", which corresponds to a fifth-degree equation and three control parameters, etc.[14].

Let us briefly summarize this variational approach with the simple case of the pleat (*cusp*) (Figure 5.5), an image name that represents the shape of the response surface of the variable u according to the value of the two control variables r and q. Its equation is: $F(u; (r, q)) = (1/4)u^4 + (1/2)u^3 r + q u = 0$, including the cancelation of its derivative $u^3 + ru + v = 0$ gives the shape variation of the potential function (number and stability of stationary states u^*) (Figure 5.6).

---

13 Wood, S., Thomas, M.B. (1999) Super-sensitivity to structure in biological models. *Proc. Roy. Soc. London B*. 266, 565–570; Fussmann, G.F., Blasius, B. (2005). Community response to enrichment is highly sensitive to model structure. *Biol. Lett*. 1, 9–12.

14 A summary presentation is given by Ekeland, I. (1977). La théorie des catastrophes. *La Recherche*. 81, 745–754.

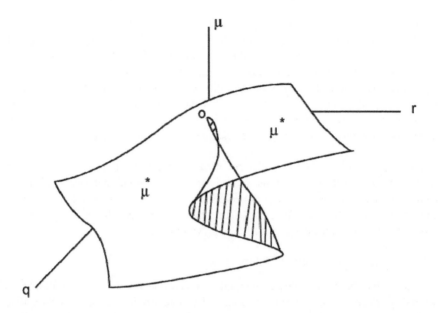

Figure 5.5. *Gathering type control surface. One state variable u and two control variables (q, r). u*: stationary states. The pleat disappears at the critical point 0*

Figure 5.6. *Variation of the potential function (vertical section in Figure 5.5). In the center: two minimum stable and one maximum unstable, corresponding to the shaded area of Figure 5.5*

5.3.5. *Multivariate structured models*

5.3.5.1. *Droop's model*

The need to compartmentalize a particular variable may be based on something other than a standard-state criterion (such as age or activity as we have seen above). In particular, we can physically distinguish, for a given substance, its external concentration in the environment, an absorbed but not yet metabolized fraction (intracellular storage) and a directly usable fraction. Droop's model (1968) was the first of its kind, developed for the study of the chemostat growth of the unicellular alga *Monochrysis lutheri*. Based on J. Monod's substrate-dependence model to study the action of a limiting nutrient s (in this case, vitamin B12, a growth-limiting factor for this phytoplankton species), it introduces a new variable q representing the concentration of this nutrient stored in the cell before its metabolism (called "intracellular quota"), s_{in} and D being the parameters for continuous renewal of the culture medium:

$$\frac{ds}{dt} = \frac{k_1 s y}{k_2 + s} + D(s_{in} - s)$$

$$\frac{dy}{dt} = \mu_{max}\left(1 - \frac{k_q}{q}\right) y - D y$$

$$\frac{dq}{dt} = \frac{k_1 s}{k_2 + s} - \mu_{max}(q - k_q)$$

Various extensions of this Droop's model allow us to take into account two or more nutrients, with applications to the study of the eutrophication of a natural environment (high content of N and P).

5.3.5.2. Models of enzymatic kinetics

Another area of application of this principle is in enzymatic kinetic models, of which here is a basic overview in the case of a chain of n coupled biochemical reactions. The regulation of such a sequential system with a loop connecting the extreme links can be described on the basis of the following two principles: (i) the final product inhibits the first reaction according to a given function $f(x_n)$ (*negative feedback* with delay determined by the number n of steps) and (ii) each of the intermediate reactions is subject to class transitions according to linear kinetics:

$$\frac{dx_1}{dt} = f(x_n) - b_1 x_1 \quad ; \quad f(x_n) > 0$$

....

$$\frac{dx_i}{dt} = k_i x_{i-1} - b_i x_i \quad i = 2,...,n$$

B.C. Goodwin's model (1965)[15] adopts for regulation function f a classical saturation function of the type:

$$f(x_n) = \frac{k_1}{1 + x_n^\rho} \quad ; \quad \rho \in \mathbb{N}^+$$

---

[15] Not to be confused with the American economist Richard M. Goodwin (work on economic growth).

Inhibition of the first reaction of the sequence is the result of the combination of p molecules of the final product x_n with the enzyme catalyzing this reaction (as a denominator of the coefficient k_1).

The value of parameter p (Hill's coefficient[16]) ensures a qualitative distinction of regime:

– for $p = 1$, the equilibrium point, locally stable, is an overall attractor for this system; there is no periodic solution;

– if $p \geq 2$, there is cooperative inhibition (sigmoid response, becoming threshold for large p).

In particular, it is shown that for $p \geq 2$ with n large enough, there may be an unstable equilibrium and oscillations. The mathematical analysis of the model (using the Laplace transform) gives us the stability equation explaining the conditions of instability and the establishment of an oscillatory regime, solutions (coefficient values) to be validated experimentally.

The principle of this type of negative feedback applies to the formulation of genetic regulation. In the elementary case of an enzymatic reaction (one enzyme x_2) controlled by the product x_3 whose formation it catalyzes, regulation is carried out by inhibition (repression) of the transcription of the gene encoding messenger RNA (x_1) according to the graph:

$$\begin{array}{ccccc} & & & & substrate \\ DNA & \to & mRNA\ (x_1) \to enzyme\ (x_2) & \to & \downarrow \\ \uparrow & & & & \downarrow \\ \leftarrow & & \leftarrow & & product\ (x_3) \end{array}$$

So follows the written dynamic system with the corresponding concentrations:

$$\frac{dx_1}{dt} = \frac{k_0}{1 + \alpha x_3^p} - b_1 x_1$$

$$\frac{dx_2}{dt} = g_1 x_1 - b_2 x_2$$

$$\frac{dx_3}{dt} = g_2 x_2 - b_3 x_3$$

---

16 Hill's equation was developed for the fixation of O by hemoglobin: $V = (y^p)/(1 + y^p)$, where y is the concentration of free ligand. The fixation curve (in fraction of sites) is a sigmoid for $p \geq 2$.

An interesting and biologically relevant modification consists in replacing the linear degradation of product *P* by the saturation function of a kinetics of Michaelis–Menten:

$$\frac{dx_3}{dt} = g_2 x_2 - \frac{k_1 x_3}{k_2 + x_3}$$

With this formulation, the model allows an oscillatory dynamic of the limit cycle type.

This classic model can be interpreted in terms of the number of active genes G_A and of repressed genes G_R in the total population of G_T genes encoding this mRNA in the sample studied:

$$G_A + \rho x_3 \underset{h_2}{\overset{h_1}{\rightleftarrows}} G_R$$

$$\frac{dG_A}{dt} = -h_1 G_A x_3^\rho + h_2 G_R$$

$$G_T = G_A + G_R = G_A \left(1 + \alpha x_3^\rho\right) \; ; \; \alpha = h_1 / h_2$$

5.3.6. *Dynamics of spatio-temporal process*

As a general rule, any process of a spatio-temporal nature requires the joint reporting of kinetic and transport equations, which should also distinguish the movement of metabolites or reactants and that of cells or individuals. In the diversity of experimental situations, only a few points will be specified in the following, in order to highlight the contribution of some essential mathematical tools to understand the dynamics of a biological process.

5.3.6.1. *Growth–diffusion–advection models*

The kernel of these models deals with both temporal kinetics (growth type) and transport action. "Transport" means either a simple movement of the substrate as a diffusion or a movement affecting the state variable. This is the case, for example, of the dynamics of natural populations whose individuals are subject to displacement (see below for so-called "spatialized" models). In addition, there is the effect of substrate structuring when the substrate has a strong topographic heterogeneity of the gradient type.

In this context, we present growth–diffusion models, the typical case of the famous Fisher equation (also known as Kolmogorov–Fisher or Kolmogorov–Petrovsky–Piskunov, named after the authors who studied it mathematically). For his part, R. Fisher focused on its use in population genetics to analyze the spread of a gene in a natural population. He considered this model as the fundamental equation of his theory of natural selection (1930). Let us remember here its formalism, which combines the principle of logistical growth and displacement by diffusion according to Fick's law. Either, in standard form for the variable C of the diffusion coefficient D:

$$\frac{\partial C}{\partial t} = \alpha C(1-C) + D\frac{\partial^2 C}{\partial x^2}$$

$\alpha > 0, D > 0$, unidirectional diffusion according to x.

This completely classical model presents some constraints on the initial conditions for obtaining a mobile growth wave, a question that has been well studied. Other considerations are also of practical importance, such as the problem of critical size in a limited environment. It should be noted that logistic growth subject to diffusion over a sufficiently large habitat results in a stable distribution of the steady state. We limit ourselves here to the presentation of some trajectories corresponding to the existence of a wave under the condition $c \geq 2\sqrt{(\alpha D)}$, explained in Figure 5.7, after changing variables: $C(x, t) = U(z)$; $z = x - ct$.

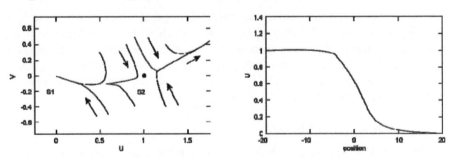

Figure 5.7. *Fisher's equation of logistic growth with diffusion. Stationary points: S1 (stable) and S2 (unstable), c = 2.2 (see text)*

5.3.6.2. *Reaction–diffusion models*

We have previously noted the contribution of mathematician and computer scientist A. Turing to morphogenesis modeling (section 3.7.2). In this field, in fact, the phenomenon studied, regardless of its physical nature, is posed as belonging to a double dimension, spatial and temporal, which formalism must explain. Any differentiation (e.g. such cellular differentiation in an even more or less totipotent tissue) must be studied from a kinetic point of view (its conditions of realization), on the one hand, and from a spatial point of view (in which site), on the other hand. Any morphogenesis is by definition concerned by this dual organization: it occurs only at a given time and in a given place. In addition, particularly in liquid environments, there are possible displacements (structuring waves) such as for certain oscillating chemical reactions (such as the famous Belousov–Zhabotinsky reaction).

These reaction–diffusion models are designed on the principle of coupling between state variables called "morphogens", whose evolution results from their interaction (as for any kinetics) and their diffusion on a given substrate. However, they are not quite field models in the sense that only the velocity equations of the morphogens Y are posed, i.e. in the case of diffusion in only one direction x:

$$\frac{\partial Y}{\partial t} = R(Y) + D_x \frac{\partial^2 Y}{\partial x^2} \qquad [5.4]$$

$R(Y)$ being the reaction functions of morphogens.

A field model *stricto sensu* (see below) would study the displacement of any point in the field as a function of both local morphogen concentration and position, which equation [5.4] does not do.

A first illustration of this model considers with A. Turing the elementary case of a ring of n indexed cells i, $(i = 1,..., n)$ and two morphogens X and Y scattering between adjacent cells. Such a physical system can be seen as a simplified analogue of certain biological structures (e.g. blastula in animal embryology or, for the primary structure of plants, the annular region delimiting the cortex and medullar parenchyma where the conductive vessels, phloem and xylem will differentiate).

Suppose that this system is close to equilibrium. The deviations from this, noted x_i and y_i, can be expressed as a first approximation by the linear system:

$$\frac{dx_i}{dt} = a x_i + b y_i + D_x \left(x_{i+1} + x_{i-1} - 2 x_i \right)$$
$$\frac{dy_i}{dt} = c x_i + d y_i + D_y \left(y_{i+1} + y_{i-1} - 2 y_i \right)$$

With this coupling of local reactions and linear diffusion, we can expect *a priori* an evolution towards uniformity of concentrations. This system can therefore only generate a spatial structuring according to conditions on the value of the parameters. Let us say, in short, that a Fourier transformation of these *x* and *y* leads to a new system, whose solutions include time exponentials affected by complex numbers. However, we know that the dynamics of a very simple linear system of dimension *2(x, y)* qualitatively depends on the properties of the coefficient matrix (more precisely on the algebraic nature of its eigenvalues, their sign and their real or complex nature). A. Turing gives a very simple theoretical example, based on a set of coupled autocatalytic reactions. By numerical simulation, it shows the possibility of generating stationary structuring waves (defined by local morphogen concentrations).

Turing's ideas have inspired various models of natural phenomena, including in biology the work of A. Gierer and H. Meinhardt, which emphasizes L. Wolpert's (1969) notion of positional information. H. Meinhardt's models are based on the catalytic coupling of two morphogens, an activator *A* and an inhibitor *H*, with very different diffusion coefficients, hence the establishment of clear concentration gradients, quite different from each other in amplitude and extent. It is the *H* inhibitor, much more diffusive, which provides the positional information function for any cell in the field thus traversed, namely:

$$\frac{\partial A}{\partial t} = \rho_A + c\frac{A^2}{H} - \mu_A A + D_A \frac{\partial^2 A}{\partial x^2}$$

$$\frac{\partial H}{\partial t} = \rho_H + c A^2 - - \mu_H H + D_H \frac{\partial^2 H}{\partial x^2} \quad ; \quad D_H > D_A$$

To this basic formalism are associated other equations according to the type of process studied. For example, for cell differentiation determined by activator *A*, two other variables are involved: *S*, of a trophic nature (subject to depletion), and *Y*, whose role is to qualitatively switch or bifurcation to the differentiated state. This implies that this variable *Y* has multistationarity. Mathematically, we have the following solution, resulting from a classical saturation mechanism (see J. Monod's substrate dependence model) limiting the synthesis of this variable *Y* according to the equation:

$$\frac{\partial Y}{\partial t} = d A - e Y + \frac{Y^2}{1 + Y^2} \qquad [5.5]$$

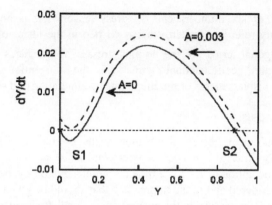

Figure 5.8. *Meinhardt model of cell differentiation. Equation[5.5]. *: stable stationary states S1 and S2*

Figure 5.8 shows how this coupling works (S, Y). In the absence of an activator ($A = 0$), there are three stationary states $dY/dt = 0$ (two stable S1 and S2, and one unstable). In the vicinity of S1 at low concentrations of Y, we have $dY/dt < 0$, so that the system evolves to S1, where $Y^* = 0$ (no differentiation). The differentiated state S2 can only be achieved for a particular initial state (high concentration of Y) and above a threshold located in this figure by the unstable stationary point between S1 and S2. *A contrario*, in the presence of A, we have $dY/dt > 0$, showing that even with a very low initial concentration of Y, there is evolution towards the differentiated state S2.

The regeneration phenomenon, so important in animal or plant morphogenesis, benefits from this reaction–diffusion approach. A good example is the regeneration of the green algae *Acetabularia*, modeled by a *mechanochemical model* (Goodwin and Trainor). This is a natural phenomenon that occurs in the spring after the winter beheading of this algae. This model differs from the previous ones in that it takes into account both the mechanical stresses and the kinetics of the calcium subjected to diffusion. The role of Ca as a morphogen is to act on the viscoelastic properties of the cytogel. Mathematically, this study uses specific mathematical tools well known in continuous media mechanics, stress and displacement *tensors*. This makes it possible to simulate the regeneration of the cap of these algae, reproducing the geometry of the apex and the periodic structure typical of the apical crown of whorls (Brière and Goodwin).

REMARK.–The notion of "tensor" generalizes that of vector, according to the following distinction of the dimensions of quantities: scalar, vector, matrix and tensor. For example, a pressure is represented by a scalar (dimension 1), a force by a vector (three numbers in R^3 in relation to the chosen base). A mechanical stress is defined by a set of six numbers called "stress tensors". It is an invariant (i.e. independent of any reference system) that is written in the form of a matrix 3×3 symmetrical. Diagonal terms are the normal stresses on the faces of the element under consideration; extradiagonal terms are the tangential stresses. They correspond to the displacements or modifications by elongation and shearing.

5.3.6.3. Field models

Since any biological object is inhomogeneous by nature, field models are intended to locally describe *a particular physiological or morphogenetic activity*[17]. The principle of the notion of field is a little ubiquitous in biology, both in the nature of the process considered and in the importance that should be given to the local. To illustrate this notion, let us turn to the exemplary and well-studied type of study of a growth field in plants, organisms in which the growth activity within a tissue or organ is generally very unevenly distributed.

This distribution concerns, on the one hand, the growth rate itself, and on the other hand, its direction (anisotropy). For example, for the growth field constituted by the expansion of a planar leaf blade, the consequences of these two aspects of tissue inhomogeneity may relate to a change in the shape considered in 2D (affecting the usual length/width allometry) and/or its local variations in thickness (embossed leaf faces). The measurement of an overall growth intensity must therefore be extended by a mathematical characterization of the growth field that can be related both to local mitotic activity (frequency and direction of mitosis) and to specific metabolic properties.

Let us present the principle of analysis of a growth field in the relatively simple case of unidirectional growth, such as the elongation of a young root. Beyond the macroscopic distinction of its different regions (meristematic, cellular elongation, maturation, branching) well known since H.L. Duhamel du Monceau in the 18th Century, the analysis of a root field did not begin until the 1940s to the 1950s on young maize roots, with the definition of a particular quantity, called "specific elementary speed of growth in length" (R.O. Erickson), which expresses the activity of an element of root length. The problem is thus posed in the context of vector analysis considering the analogy between the growth of a tissue or organ with the dynamics of a continuous medium.

The calculation of the elementary specific growth rate in length is based on the kinematics of markers or particles deposited on the root surface, whose displacement

---

17 The growth field considered here has nothing to do with the term "morpho-genetic field" (or "morphic") used in a completely different context and especially without the precision provided by the mathematical treatment presented in this chapter based on the definition of duly measurable field variables.

is monitored during organ elongation. Let x be the position of a given material point M (measured with respect to the root tip O) and v the growth velocity of the OM segment. Knowing that the growth of an element of root length dx depends on its position x and time t, the fundamental equation of local growth is written:

$$dv = \frac{\partial v}{\partial x} dx + \frac{\partial v}{\partial t} dt$$

In practice, in its usual use, it is generally assumed that it is stationary. The field of growth velocity is assumed to be time-independent: $\partial v / \partial t = 0$. Under these conditions, the specific rate of elemental growth is defined as the divergence of the velocity vector:

$$v_{élém} = \frac{d}{dx}\left(\frac{dx}{dt}\right) = \frac{dv}{dt} = div(\mathrm{v})$$

The mathematical treatment of this notion of the rate of growth of an infinitesimal element makes it possible to go further, with the possibility, as in the dynamics of continuous environments, of a double physical representation, called Eulerian and Lagrangian, which provides additional information on the growth process.

The Eulerian or spatial representation corresponds to the function $v(x|t\ fixed)$, experimentally estimated by photographic recording of the displacement of the markers. Its derivative $(\partial v / \partial t)_x$ has the meaning of acceleration. With the Lagrangian or material representation, we are interested, not in what happens at the position x on the organ, but to a given material element that has been visualized by a particle M deposited on the surface of the organ. From the displacement of this "particle" (i.e. of a particular cell or group of cells), the Lagrangian derivative for the point M, write:

$$\frac{Dv}{Dt}(M) = v_M(x_{OM}, t) = \frac{\partial v}{\partial t}(x) + v\frac{\partial v}{\partial x}$$

In this expression, the last member distinguishes between local changes at position x and the displacement of point M. The use of the Lagrangian description is justified from a biological point of view, because the interpretation of the analysis of a physiological field is based on what happens at such a tissue site (say such a cell that ideally should be followed) rather than at a point fixed by its geometric position and which is not occupied by the same cell.

In 2D growth, for example, the expansion of a leaf blade, also uses the previous vector analysis. Let us summarize the developments it brings. On the one hand, it allows a mapping of local growth activity and its variation during ontogenesis, in particular, the identification of the displacement of the fastest growing region. On the other hand, it makes it possible to express *growth anisotropy* and, correlatively, its variation during the organ growth. In addition, the variation in direction of the highest elongation, or vorticity, is measured, by analogy with the rotational motion of a fluid particle, by the vortex or rotational vector of the growth vector[18], to which we associate a *vorticity tensor*.

Finally, let us note the importance of certain 3D growth fields. Thus, in the functioning of apical meristems, an essential role is assigned to the privileged orientation of mitoses (called anticlines or periclines according to their direction perpendicular or parallel to the surface of the meristem). The field analysis highlights a fundamental notion that is the existence of principal growth axes (Figure 5.9). The main point of this study (Z. Hejnowicz) applied to a meristem is to show that cells divide according to the principal planes defined by the growth tensor. Let us add, in the analysis of the dynamics of these meristems, the use of a curvilinear coordinate system, called "natural system in accordance with the geometry of the apical dome" (e.g. paraboloidal system).

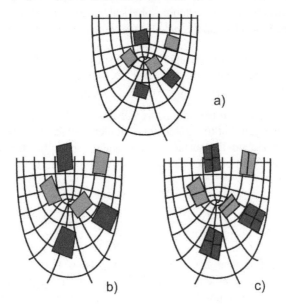

Figure 5.9. *Root apex (simplified scheme after Nakielski and Hejnowicz (2003)). a): Initial state of cells located in the various quadrilaterals specified in different arrangements. The deformation of these quadrilaterals indicates the direction of the principal growth axes. b): Cell elongation without division. c): Growth with division. For a color version of this figure, see www.iste.co.uk/buis/biology.zip*

---

18 This rotational is the vector $\omega = \text{rot}(v) = \nabla \wedge v$, where $\wedge$ refers to the vector product.

Another important aspect of this concept of a growth field is that local distribution can have dynamic properties that modify the field facies during tissue ontogeny. This is because it is not only enough to describe local variations in inhomogeneity, but also to specify their relative stability. We know that if a given plant tissue can be seen as a "patchwork" of micro-regions of very diverse activities, their distribution can be subject to remarkable oscillations. In short, let us say that each of these micro-regions is likely to cyclically evolve from a state of high activity to a state of inhibition or latency. Such cases, known as "mosaic growth", can be observed in plant cells where it has been clearly demonstrated that the wall extension is in the form of a hypercycle, i.e. a type of loop network coordinating spatio-temporally different elementary components (multi-enzymatic processes under the dependence of the electrostatic state of the wall, incorporation of new parietal elements) (see Figure 3.11).

5.3.6.4. *The notion of hypercycle*

With these new models, the position variable is not explicit. The spatial component intervenes in a completely different way, in the form of external inputs within a cycle of autocatalytic processes. These inputs are specified by their point of impact on the cycle, i.e. on the progress of the process. There are in a way two types of effectors; on the one hand, the inputs that correspond to the dependence on a substrate associated with the studied process, and on the other hand, the evolution within a series of reactions until the achievement of an anticipated *ad hoc* state. For example, the determinism of plant cell growth here jointly results from the electrostatic state (ionic charge) of the cell wall, the incorporation of new materials (wall extension) and the development of a multi-enzymatic process. Such a model makes it possible to resolve the contradiction between the properties of enzymes deduced from *in vitro* studies and their effective behavior *in situ* (fixed enzymes, parietal electrical potential). We refer to section 3.11.2 that presents the overall regulatory scheme (Figure 3.11).

This review of different types of continuous models is far from exhaustive. For example, all biomechanical models involving one or more mechanical variables (turgidity, particularly important water potential in plant physiology) and where specific quantities (stress or strain tensors) must be taken into account have been left out.

5.3.7. *Multi-scale models*

This elaborate form of structured models has quite recently developed from the first studies that focused on material stability. It also has its place in the analysis of living systems, as soon as we want to take into account in some detail the fact that there is indeed a multi-scale (or multi-level) organization in many biological processes. Indeed, many phenomena often occur or result from the coordinated intervention of several organizational levels that interact, from the molecule to the organism and the ecosystem. See the overview of A. Lesne (Lesne 2009) on the multi-scale organization of living systems.

It is no longer just a question of structuring a set of variables as when we establish classes or categories of cells according to their state (age, activity classes, etc.), whose interactions and respective evolutions would simply have to be analyzed. Other levels of functioning that may be involved in the process are added. Example: the bonds between the molecular level and the cellular level, any molecule or, more precisely, any molecular network or cycle, acting on a certain cellular site with which it interacts.

The principle of this approach can be summarized in the following simplified way[19]. Let us consider two kinds of variables and their related spaces:

– a macroscopic variable N in a space x (physical space where the cells are located);

– a microscopic variable n that represents a set of cellular states in a space between which we place a system of equations describing their respective dynamics:

- macroscopic dynamics on x: $F_x(N, n) = 0$;

- microscopic dynamics on y: $f_y(N, n, x) = 0$.

As an example of interpretation, let us say that $N = N(x, t)$ can represent the cell density on x, and $n = n(x, t; y)$, the distribution of cellular states y at a given x-position. We see the application of this principle and its extension to the study of cancerous growth, which depends on several determinants: genetic transformations, growth of already individualized tumors, interactions with the body and the environment. Similarly, this principle can be applied to any other situation, such as in ecology, where the importance of a chain or network of different levels is essential in the dynamics of a particular type of process.

From this point of view, it is clear that most of the dynamic systems used in biology can be seen as approximations that, for one reason or another, neglect the reality of this organizing principle. Even reaction–diffusion systems, such as those well-documented by A. Turing (morphogenesis with emergence of spatial structuring) and Kolmogorov–Petrovskii–Fisher (gene dynamics in a population), are, if we can put it that way, only relatively simplified tools, as they do not take into account the interplay between organizational levels. The only interactions they take

---

19 We are inspired by the presentation given by Bernard, S. (2013). Modélisation multi-échelles en biologie. In *Le Vivant discret et continu*, Glade, N., Stephanou, A. (eds.). Éditions matériologiques, Paris, 65 *sq*.

into account are all at the same level. With A. Turing, for example, we consider the interactions at the molecular level of morphogens according to a classical kinetics, the cellular level being apprehended only by the role played by positional information. It is not the addition of the genetic regulation level (Meinhardt 1978, 1982) that in itself can be sufficient to talk about a multi-scale model, as their interweaving should be clarified. Another example of complexity is the modeling of a chemotaxis phenomenon where two types of transport are considered: the movement of cells and the diffusion of substances (nutrients or others). We then base ourselves in a completely classical way on their own kinetics (formation, degradation, consumption) and on their transport, but neglect the interaction between these two interdependent levels, namely the cellular population and molecular distribution. However, of course, the practical choice of the type of model depends on the objective pursued, the development of a highly integrated model involving on the one hand referring to a corpus of sufficiently explicit hypotheses and, on the other hand, having experimental possibilities for validation, not to mention the mathematical solution of equations quickly becoming rather complicated (especially in the case of partial derivatives).

An example of a multi-scale approach is the cell dynamics model of V. Volpert et al (2013), applied in particular to the regulation of erythropoiesis[20]. The generation of red blood cells (erythrocytes) in the bone marrow and their fate are subject to two types of regulation at two distinct levels. Without detailing the mathematical formalism used, let us summarize the principle of the elementary process involved, which it is necessary to jointly insert into a system of differential equations. First of all, it is necessary to consider the essential importance of intracellular regulation with the key role of two protein precursors (called ERK and Fas) in a competitive state, hence a bistable dynamic that improves the choice (bifurcation) between very different behaviors: either proliferation without differentiation or differentiation and then mortality (apoptosis). Second level to consider, it is necessary to add the existence of an extracellular regulation where different compounds can play, but which can be modeled by a reaction–diffusion-type process with a medium-dependent particle movement (referred to Darcy's physical law). Let us remember with this example the methodological evolution that it entails in relation to Droop's model already mentioned (section 5.3.5.1), which was simply based on an elementary structure distinguishing three levels, the extracellular fraction, an intracellular fraction absorbed and stored on standby, and the directly usable metabolized compartment. These were then simple categories and not real organizational levels characterized by differences in time scales.

---

20 Volpert, V., Bessonov, N., Eymard, N., Tosenberger, A. (2013). *Modèle multi-échelles de la dynamique cellulaire*. In *Le Vivant discret et continu*, Glade, N., Stephanou, A. (eds.). Éditions matériologiques, Paris, 91–111.

5.4. Discreet formalism

Let us recall the introduction of a discreet formalism in biology with the mathematical *Fibonacci sequence*:

$$u_n = u_{n-1} + u_{n-2}$$

This prototype of linear recurrence equations was given as a representation of a biological growth phenomenon on which we specified the strong hypotheses that could justify its use. A first extension was considered by modifying Fibonacci time lags 1 and 2:

$$n_t = n_{t-m} + n_{t-n} \; ; \; m \neq n$$

This sequence applies to the development of an algal filament, with n being the number of cells of which there are two categories according to their lifespan m or n (i.e. their division times, fixed according to their position at birth, such as their polarity with respect to the direction of development of the filament).

Another generalization leads to the principle of well-known *autoregressive models* for the study of time series, which can be complicated in various ways with the abandonment of the linearity hypothesis:

$$x_{i,t} = f_i\left(x_{i,t-h}, x_{j,t-h'}\right)$$

and the introduction of a delay h, which can be multiple.

The interest of a discrete formalism concerns in particular the various models inspired by the Automata theory, in particular the L-systems whose application we have seen to model a morphogenesis. Another well-specified case is that of linear models, known as *matrix models*, widely used in structured population dynamics:

$$n(t) = M\, n(t-1)$$

with n being the vector of the numbers of the different state classes (age, for example) and **M** the transition or projection matrix. In the case of a structuring in age groups, we speak of "Leslie's matrix", referring to the pioneering work of this author (1945, 1948). Let us underline the hypotheses: linearity of relationships, no memory or delay effect.

These matrix models (Caswell 1989) are to be compared to Markov processes, which are random processes where the probability of occurrence of a state depends only on the probability of the previous state.

5.5. Spatialized models

Two very different types of spatialized processes are to be considered. We have seen a first type with Turing's reaction–diffusion systems, then with field models, both corresponding to the study of phenomena of a spatio-temporal nature by studying local singularities. We take up this question here from two other points of view. On the one hand, it is a question of introducing the idea that these are often **collective phenomena** involving an interest in individual behavior and not only in average values. This is the purpose of some multi-agent models, explicitly referred to as "**individual-centered models**". On the other hand, we have another category of models based not on the dynamics of usual statue variables (such as numbers or concentrations) but also on the formation and displacement of an electrical potential (**electrophysiological processes**). We will give an overview showing their mathematical particularities, compared to the models previously presented.

5.5.1. *Multi-agent models: dynamics of a biological association of the individual-centered type*

We have seen that the spatial representation of a biological process can be studied using a dynamic system with one or more diffusion and/or advection terms. The remarkable case of Turing reaction–diffusion systems thus allows the study of the spatial distribution of state variables with, in particular, the existence of stationary displacement front and the formation of a structure. In these classical models, space is considered as a field of concentration or density related to state variables without making any material reference to the "individual" themselves (with the exception, however, of the so-called material or Lagrangian representation of a univariate growth field).

There is another approach that is particularly suitable for biological associations. Unlike the classic formulations of A.J. Lotka and V. Volterra, which give an "average" view of the behavior of each species, it is important to be interested in the fact that these are **collective phenomena** that involve taking into account locally, on the basis of state variables, the dynamics of individuals. These are simply distinguished by their membership in a particular group, each group having its own spatial and temporal kinetics. This is the case, for example, of multi-species models of association of competing species with the objective of detailing the spatial distribution of each agent (as such species) in an evolving inhomogeneous field.

This spatialization problem concerns various processes at different organizational levels, from cellular to ecosystem. For example, in ecology, the identification and monitoring of areas of territory mainly devoted to a certain group (regionalization or patchiness). The method consists in monitoring the movement of any individual according to their own growth and local interactions with their surroundings (of the same species or not), hence the name **individual-centered models**.

From a practical point of view, this approach is, therefore, not limited to the classical dynamics of state variables alone but involves a simulation of the probabilistic behavior of any individual. Computer programs, such as NetLogo, NetBioDyn or Stella, meet this objective of a dynamic spatialized representation for collective phenomena whose nature can be very varied. Figure 5.10 provides a simple example for a predator–prey model applied to the association of two vegetation–herbivore species, whose growth rates are determined, with the multiplication of the herbivore being determined by the level of its consumption.

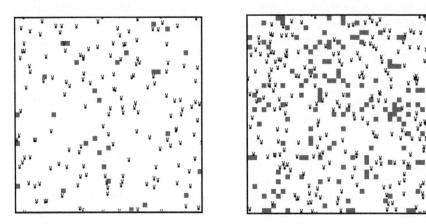

Figure 5.10. *Individual-centered predation model of the rabbit–grass association. Simulation by NetLogo software. The rectangles represent an area of grass development. Left: initial stand; right: evolution after 200 iterations. For a color version of this figure, see www.iste.co.uk/buis/biology.zip*

5.5.2. *Electrophysiological models: transmission of electrical signals*

With the first evidence of electrical activity in animals by L. Galvani in the 18th Century, the analogy was made between "nervous fluid" and "electrical fluid". Later, the first electrocardiogram recordings came at the beginning of the 20th Century. However, it was only in the 1930s that this new field of physiology, electrophysiology, was really born, with the notion of membrane potential, the measurement of which became possible through the use of microelectrodes.

Referring to Nernst's electrochemical law for a given ionic compound (e.g. K), the membrane potential resulting from a difference between intra- and extracellular ionic concentrations can be expressed by the relationship:

$$V = k \ln\left([K]_e / [K]_i\right)$$

where the potential V is in volts. Parameter k depends on the number of electrons exchanged, the temperature (in Kelvin) and the constants of the perfect gases and of Mr. Faraday.

On this particular subject, we can briefly recall some basic elements of electrophysiology (well summarized, for example, in (Françoise 2005)). It is known that, in the case of an excitable element (neuron), the effect of an electrical stimulation is to cause locally, beyond a threshold of excitability and according to an all-or-nothing law, a sudden and transient inversion of its resting potential (depolarization), then its repolarization with the appearance of an action potential (nervous influx) that propagates without modification along the excited element. By analogy with an electrical circuit of intensity I comprising in series a capacitance C and a resistance R, the electrical potential $v(t)$ of a neuron is determined by the equation:

$$\tau \frac{dv}{dt} = v(t) + RI(t)$$

where $\tau = RC$ is a time constant.

Writing for the threshold value: $v_s = v(t_f)$, we have the following expression of the potential:

$$v(t) = v_s \exp\left(\frac{t - t_f}{\tau}\right) - \int_0^t A \exp\left(\frac{t - s}{\tau}\right) I(s)\, ds$$

For $t < t_f$ (integral zero), the potential increases by an initial value $v_0 = v_s \exp(-t_f/\tau)$ to v_s, then decreases sharply before resuming exponentially. The effect of a generating pulse can be represented by a sum of functions of P. Dirac, resulting in the characteristic shape of a series of peaks of action potentials.

The basic mathematical model of electrophysiology is the **Hodgkin–Huxley model**, experimentally developed on the giant axon of squid (1952). Its formulation is based on the transmembrane currents of the Na^+ and K^+ ions, responsible for depolarization and repolarization of the axon, respectively, according to the electric charge balance equation:

$$I = C_m \frac{dV}{dt} + I_{Na} + I_K + I_f$$

with I_x representing the different ionic currents (sodium, potassium and other currents noted f for leakage current). The model state variables here are not concentrations, but ion activation functions: K *n(t)* activation, Na *m(t)* activation and Na *h(t)* inactivation functions. Finally, taking into account some hypotheses that are not necessary to detail here, T. Hodgkin and J. Huxley establish their relationships with the equilibrium potentials for each ion according to Nernst's law and the corresponding membrane conductances. In addition to these three differential equations on time, there is the equation of the resulting action potential *dv/dt*.

The **FitzHugh–Nagumo model** is a simplification of the Hodgkin–Huxley equations, taking into account the important fact of a difference in scale between the rapid activation of sodium and the slower electrical response. The system is reduced to two relatively simple equations that can be written according to new variables *x* and *y*: on the one hand, *dv/dt* of the fourth degree in *n* and the third degree in *m*, and on the other hand, *dn/dt* linear in *n*. With a change of variables, this model becomes:

$$\frac{dx}{dt} = k\left(-y + 4x - x^3\right)$$
$$\frac{dy}{dt} = x - by - c$$

5.6. Random processes in biology

It is known that the introduction of a random term into a velocity equation can significantly alter dynamic behavior. Thus, in population dynamics, the deterministic predation model of Lotka–Volterra provides for an oscillatory dynamic whose temporal characteristics are determined by the initial conditions with the coexistence of the two species maintained. This model is sometimes presented as a basis on which to make it more realistic by adding a random component. However, in fact, this can lead to a change in the dynamic regime that can lead to the extinction of one of the species. Technically, this can be easily obtained using a Langevin equation of the type $dy/dt = f(y) + g(y)\xi(t)$, $\xi(t)$ representing "white noise" (a time-independent Gaussian variable with no expectation and a given variance).

However, such a modification carried out by simulation[21] does not allow, as is sometimes written, us to give this addition the meaning of an interaction between organizational levels. It is simply the introduction of a Brownian motion transport component, not a multi-scale game. This remark means the risk of confusion in the statement of assumptions preliminary to any modeling, which are supposed to support a biological interpretation, otherwise model simulation exercises lose their explanatory value.

The *raison d'être* for this type of methodology is the great variability of biological quantities. Let us present the consideration of this property by referring to one of its important fields of application, the dynamics of a population, whether cells or higher organisms. Even when a population has fairly precise global characteristics by closely monitoring, for example, a deterministic growth law (such as a logistics for sigmoid kinetics), it is by nature made up of an inhomogeneous set of individuals. For example, it is known that a cell population in *in vitro* culture, under experimentally well-controlled conditions, always exhibits a high variability in the lifetime of the different cells. There is never a rigorous synchronization of cell divisions, even from an initial implant that is as homogeneous as possible. It is on this double observation that an approach by random (or stochastic) processes is justified, the principle of which is as follows.

PRINCIPLE.– From the random behavior of an individual (microscopic level), we try to deduce the behavior of the population (macroscopic level), these two levels being characterized by different scales, dimensions and time.

"Stochastic process" refers to the evolution of one or more time-dependent random variables t: $\{X(t)\}$, variables and times can be continuous or discrete. These are models with a good degree of generality, regardless of the nature of the population (gene, cell, organism, sub-population), which also apply to processes other than demographic change.

As the evolution of the population size is basically a balance between birth rate and death rate, the phenomenon emerges from a *birth–death process*. Let us give an overview in the simple case of Markovian behavior. In this context, a **Markov process** is defined as follows: an individual is born or dies regardless of what has previously happened and regardless of the behavior of other individuals in the population. In other words, being a process without memory, adding information

---

21 Bernard, S. (2013). *Modélisation multi-échelles en biologie*. In *Le Vivant discret et continu*, Glade, N., Stephanou, A. (ed.). Éditions matériologiques, Paris, 77.

about previous states is useless, which is mathematically written in terms of conditional probabilities:

$$P(X_{n+1} = x \mid X_0, ..., X_n) = P(X_{n+1} = x \mid X_n)$$

The state of the population at the time n is:

$$X_n = X_0 + \sum_{i=1}^{n} Z_i$$

with Z_i being independent random variables and of the same law:

$$P(Z_i = 1) = P(Z_i = -1) = 1/2$$

We have a formal analogy with random walking or Brownian motion.

The first work on this question was that of I.J. Bienaymé (1845) on the probability of extinction of a family name (probability that a man still has descendants bearing his surname after n generations). Long forgotten, this genealogical subject was taken up again in 1873 by F. Galton and H.W. Watson. It is because of the diagram graphically representing the sequence of filiations that these studies were given the name "*branching process*" (or "ramification process") or "tree" or "Galton–Watson process". Such a process is simply a random sequence of generation-by-generation numbers (Figure 5.11). If we assume that the individual probability of having k descendants in each generation is p_k, this basic process is equivalent to a Markov chain. By extension, this idea was adopted for other subjects, such as population genetics (estimation of a recombination rate), species evolution (already by F. Galton who was a cousin of C. Darwin) or epidemiology.

The basic mathematical tool is the density functions, the integral of which (distribution function) gives the demographic evolution of a population. In reality, density functions are multivariate in nature, because it is necessary to take into account this or that criterion of continuous population structuring. For example, the cell density of the size is written $n(t,x)$. An extension of this principle of temporal connection consists in what is called a spatial connection for which the reproduction law (probability p_k of the number of descendants) is associated with a topographic dispersion law.

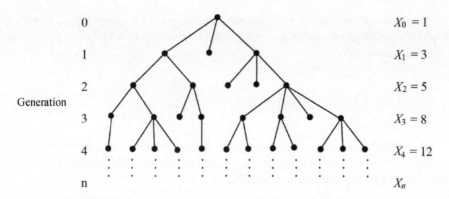

Figure 5.11. *Diagram of a simple time connection process*

Without going back to the general principle of time-series autoregressive models, let us quickly see how the idea of a random process takes place in biology, extending what Leslie's matrix models offer us in a deterministic version for the population growth number of a population.

5.6.1. *Poisson process*

This archetype of Markovian random process is a strict birth process (without mortality) in continuous time. It studies the occurrence times of a given event (birth) under the following hypotheses: (i) the probability of occurrence in a small time interval is proportional to the length of that interval according to a proportionality coefficient λ; (ii) independence of the number of occurrences in disjointed intervals (= independent increases); (iii) the probability of more than one occurrence in a small interval is assumed to be zero or negligible. Such processes are said to be "homogeneous over time", the increase does not depend on time. We show that, under these conditions, the probability of k occurrences at time t is given by Poisson's law of parameter λ:

$$P(N(t) = k) = \exp(-\lambda t)\frac{(\lambda t)^k}{k!} \ ; \ k \in \mathbb{N}^+$$

The interval between two successive occurrences (or waiting times) is thus a random variable obeying this law. Either the probability of a queue of j elements between instants t and t':

$$P[N(t)-N(t')=j] = \exp(-\lambda(t-t'))\frac{(\lambda(t-t'))^j}{j!}$$

The expectation of a Poisson variable being equal to the parameter λ, this counting process shows a linear growth trend: $E[N(t)] = \lambda t$. The parameter λ has the meaning of a birth speed coefficient: $dN(t)/dt$.

REMARK.– With this counting Poisson process, the probability that a cell will divide into an interval Δt is $\lambda \Delta t$. The corresponding birth process must take into account that the probability of concomitant division of several cells over the entire population is $\lambda N(t)\Delta t$. It is, therefore, no longer homogeneous over time.

5.6.2. *Birth–death processes*

A death process can be defined in a similar way to the previous one by generally assuming that the distribution of lifetimes follows an exponential law of parameter μ. The probability of life of any cell being $\exp(-\mu t) = p(t)$, the distribution function $N(t)$ approximately follows a binomial law $N(t) \approx \mathcal{B}(N_0, p(t))$ whose expectancy is: $E(N(t)) = N_0 \exp(-\mu t)$.

A simple birth–death process can be established on these bases, combining the two previous processes. These processes are of course only approximations that we give here as basic examples. It must be considered that the functions of birth rate and mortality are generally density-dependent: $\lambda(N)$ et $\mu(N)$. For example, we can have a term for logistical birth braking: $\lambda(1 - N/N_{max})$. In addition, there are certain decisive kinetic criteria (affecting birth rate and/or mortality), not to mention the existence of correlations between daughter cells and mother cells involving a memory effect (contrary to previous Markovian behavior). Thus, different models are available depending on the multivariable nature of the *N(t)* cell density function. The structuring criterion that differentiates them can be, for example, by referring to the most classical models, age (Sharpe–Lotka and McKendrick–von Foerster models) or maturation rate (Rubinov, Frenzen–Murray).

Let us clarify a little the handling of these density functions with the example of McKendrick–von Foerster's model of structuring on an age density $n(t, a)$. Let us look at the evolution of the age class a in terms of cell density:

$$n(t+dt, a+dt)\,da - n(t,a)\,da = \lambda(a)n(t,a)\,da\,dt$$

where $\lambda = \lambda(a)$ is a simple exit coefficient from age group a (and not from birth). By serial Taylor development in the vicinity of (t, a), we obtain the following partial differential equation:

$$\frac{\partial n}{\partial t} + \frac{\partial n}{\partial a} = -\lambda n(t,a)$$

which is the conservation equation for the number of cells in age group a. From its integration (theoretically up to ∞, where $n(t, \infty) = 0$), we obtain the evolution of the age group a:

$$\frac{dN}{dt} = \alpha(t) - \int_0^\infty \lambda n(t,a)\,da$$

where $\alpha(t)$ represents entries into age class a (actually by maturation from other classes). The model also formalizes the birth process (B of birth) according to the equation (called "renewal"):

$$B(t) = n(t,0) = \int_0^\infty \beta((a)) n(t,a)\,da$$

where $\beta(a)$ expresses the age dependency of the birth rate.

With this presentation, which is obviously only a very simplified view of a random process of population growth, we see the complexity of the problem and its approach through the principle of random processes. Two essential points of this formalization should be highlighted: (i) the choice between several density criteria (age, maturation, size) and (ii) the variation of these criteria during the process itself. As the problem becomes very complicated, it should be noted that we have approached solutions when we look at the long term where standard models lose their quality of adequacy (see, for example, Segel 1980).

5.7. Logic kinetics of regulation

The classical formalism of differential dynamic systems, with its search for singularities and their stability, is not the only methodology suitable for the analysis of a biological process. An alternative was developed in the 1970s by R. Thomas and S. A. Kauffman, consisting of a *Boolean approach to biological regulation circuits* (following C. Shannon, who himself used G. Boole's algebra in his communication theory).

The principle of a logical kinetics is to set a set of logical rules assigned to Boolean variables (of value 0 or 1) to replace the usual differential equations describing the kinetics of continuous state variables. This discreet formalism differs from the practice of automata theory. This logical kinetics aims to provide a *representation of the regulation process* (which is not the case with A. Lindenmayer's L-systems). We are interested in any change from + to - or vice versa, rather than simply tracking the successive values of each variable. This is particularly appropriate in cases of complex networks that are intended to be described in this way in a refined manner, especially when precise quantitative kinetic data are not available. In short, let us say that the usual notion of state variables (such as a concentration or population density) is associated with a function describing the meaning of their variation. With these pairs of variables and functions considered together as basic elements, we are as close as possible to the exercise of regulation, since we highlight not numerical values, but their direction of variation. A graphical representation symbolizes this approach, where nodes are the elements at play and arcs the regulations.

The simplest configuration is that of a monomolecular autocatalysis. Let us consider the variable α, with its variation noted here $a = \alpha$ according to the kinetics:

$$\alpha \xrightarrow{a} \alpha$$

The correspondence matrix:

Variable α	Function a
0	0
1	1

Table 5.1. *Monomolecular autocatalysis. Logical rules of transformations*

means that in the presence of α (α = 1) there is continuous growth, i.e. positive *feedback* behavior triggering a *runaway* (no regulation).

The current development of these Boolean methods focuses on genetic regulatory networks. To give the principle (Figure 5.12), let us consider an elementary case where each gene is associated with a variable S (1: active gene; 0: mutated gene) and a function s (on/off gene). The product is considered as a memory variable noted σ, which records the value of s. See Figure 5.12b, which shows the system as:

$$s = S\bar{\alpha}$$
$$a = A\sigma$$

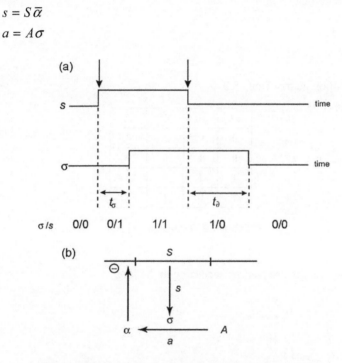

Figure 5.12. *Logical kinetics. Schematic representation of the regulation of a gene (Thomas 1978)*

For a more detailed illustration (Thomas and D'Ari 1989), let us consider the naive logical description of genetic regulation in the simple case of two linked genes X and Y whose respective products are noted x and y. Product x activates gene Y, whose product y represses gene X:

$$x \underset{-}{\overset{+}{\rightleftarrows}} y$$

By posing:

$X = 1$ if and only if $y = 0$ (X if and only if y is absent)

$Y = 1$ if and only if $x = 1$ (Y if and only if x is present)

the system is written[22]:

$X = \bar{y}$
$Y = x$

with as transition matrix Table 5.2.

x	y	X	Y
0	0	1	0
0	1	0	0
1	1	0	1
1	0	1	1

Table 5.2. *Transition matrix*

This gives the time sequences according to the loop:

$$00 \xrightarrow{x} 10$$
$$\bar{y}\uparrow \qquad \downarrow y$$
$$0\bar{1} \xleftarrow[\bar{x}]{} \bar{1}1$$

As another example of logical regulation, consider the dynamics of biological prey–predator association as described by Lotka–Volterra's classical system (Richelle in Thomas, 1979). In its simplest Boolean version, each population is assigned the following pair: a variable describing the population density (prey α, predator β) and a function (a, b) describing its variation +/- (growth or decrease) according to whether the density is below or above a given threshold. The kinetic pattern of this biological association is a retroactive loop:

$$\alpha \underset{-}{\overset{+}{\rightleftarrows}} \beta$$

---

22 The line above $\bar{y}$ means the logical complement to y (NO y).

Hence, the following chart:

$$\begin{array}{ccc} \bar{0}0 & \rightarrow & 1\bar{0} \\ \uparrow & & \downarrow \\ 0\bar{1} & \leftarrow & \bar{1}1 \end{array}$$

showing stable cyclic behavior of this biological association. This corresponds to the directions of variation when crossing isoclines in the classical phase portrait of the Lotka–Volterra dynamic system, where the trajectories are closed curves, i.e. self-sustaining oscillations (whose period and amplitude depend on the initial conditions).

This representation can be detailed to some extent by considering several Boolean variables to describe the density of each population. Remaining in a still very simple approach, a first extension would be to describe the density of each population by two Boolean (00, 01, 11) according to Table 5.3.

	Variable (density)	Function (variation)
(0 0)	Absence	No growth
(0 1)	Some individuals	Slow growth
(1 1)	Many individuals	Rapid growth

Table 5.3. *Logical kinetics of growth regulation. Population density described by two Boolean variables*

Conclusion

The panorama, the main lines of which we have just outlined on the relationships that have been woven between biology and mathematics, shows that the evolution of ideas has been a journey, not a chaotic one, but rather a marked one through a series of important stages, often intertwined or sometimes remaining in a state of latency. Thus, some innovative ideas may retain a level of abstraction with little or still no experimental validation. On the other hand, important innovations emerged or were taken up and contributed, more or less quickly, to the fact that biology began to reconsider a certain number of principles and methods. As in other disciplines, these were changes of point of view that modified the way in which we question ourselves, often very profoundly, in order to address the various specific problems of biology. This question of the *specificity of biology* as a field of knowledge and investigation remains an essential point that is still under debate. Posing this problem at the epistemological level where it should be placed, the question remains open about what a living being is. On this point, the physicist prefers to speak of the singularity of life, while the philosopher focuses more on trying to specify or define what life is.

Some figures are particularly emblematic of this evolution. We make a choice below, certainly a little subjective, based on their position and not on their real contribution to biomathematics. There is nothing paradoxical about this. Better still, it is interesting to note that innovation often results more from an idea or concept than from the systematic implementation of a properly constructed mathematical method. Hence, we would like to mention, first of all, the case of G.L. Buffon, a typical example of a naturalist who is very enamored with mathematics. Despite his pronounced taste for mathematics, G.L. Buffon remained marked by strong doubts about both the validity and operational effectiveness of mathematics in biology. Better still, he was convinced of an opposition, which he considered irreducible, between the properties of the observed real being and those of the abstract being that could be deduced from a mathematical approach, which we can specify today by saying that a mathematical equation is only an image, a formal structure posed at best as an isomorph of the real. Nevertheless, we cannot forget that G.L. Buffon had some very relevant insights. Thus, he was led to distinguish (without the help of any formalism) the external form of an organism and the existence of an "internal mold", which, in his time, could only be a simple hypothesis, or even free speculation. The significance accorded to this dual principle (internal mold + exterior directly measurable) is equivalent to a slight announcement of a new viewpoint, both topological and multiscaling.

Another essential and less paradoxical figure was that of physiologist C. Bernard, who clearly established the development of mathematical laws as a fundamental objective of his discipline. This claimed association between experimentation and formalization was historically, at least from an epistemological point of view, one of the strong points of the connections between biology and mathematics, especially as it was beyond any possible use of the mathematical methods then available. This position of principle, accepted today, was already observed in a remarkable way with Aristotle, again despite an unfavorable mathematical context, dominated at the time by the Euclidean metric that could not suit him. Indeed, it was without a mathematical apparatus that Aristotle implicitly expressed the fundamental principle of what we call an "ago-antagonist couple", clearly illustrated in the combination of acquired growth and remaining growth to be achieved, i.e. more generally, the existence linked, on the one hand, to the current state of a system (as a consequence of its initial conditions and its history), and, on the other hand, to what remains of its virtual potentialities. Another essential contribution of Aristotle was of course his concept of form (of a nature that is not reducible to geometry alone), or more precisely of the substance/formative principle couple, or, if we prefer, of a structure that individualizes itself within an amorphous or well-balanced substrate, which predicted A. Turing's famous symmetry breaking.

Mathematics and the perception of the singularities of the living

In our presentation of the connections between biology and mathematics, we stressed the primary purpose of any formalization, which is to *highlight the properties* of the biological system under study, particularly the existence of singularities. Admittedly, this objective is not always highlighted, some mathematical models being rather focused on the quality of the representation to ensure a good phenomenological reconstruction of the process under study. Graphical imagery and simulation can then be the reason behind the use of mathematics and computer science. Our conclusion would therefore be to insist on what this outlook may leave out.

We will agree that the objective of the use of mathematics is, first of all, to promote the perception of the object or phenomenon, which can thus be reduced to what we consider to be most essential at the scale considered, in the image of a form detached from an undifferentiated background. A phenomenon is best perceived when it is revealed by the description of its spatiotemporal singularities (extremums of a function representative of such a process, possibly discontinuities of them) that necessarily have a double meaning, mathematical and biological.

On this point, it should be recalled that the mathematical tool goes well beyond the immediate particularities that the biologist can detect if it is limited to the mere consideration of the extremes of a function. What matters, in fact, is not only the occurrence of minimums or maximums (of which L. Euler was very fond) in the monitoring of a process, but also the nature of the states accompanying these extremes (which is of great interest to logical kinetics). The question is to specify which is (via higher-order derivatives) the direction of variation of a given variable, especially if it is in the acceleration or deceleration phase. For example, for the same extreme of the growth rate, it is necessary to distinguish between the anterior and posterior states of this extreme, states that do not have the same biological significance since they correspond either to a stimulation phase or to an inhibition phase of the process. This *calculation of variations* (classic objective in mathematical analysis) is in line with the essential question posed by any process trajectory, which is to look at the above, even if we do not explicitly assume the existence of delay effects. The importance of a memory effect in the sense of the path taken is well-known with *hysteresis* behaviors, whose importance in various fields is now recognized in biology.

This enterprise leads to or passes through what is known as a "*law*" or a "*model*". Our present review thus considers what could eventually be considered as the bases or prolegomena of a theory, with relevant concepts. Within the limit we have set for ourselves, i.e. without naively pretending to talk about theory itself, we have been able to underline the importance of certain concepts or themes according to the particular way in which mathematics describes or translates what it is studying. This is, of course, marked by the comparison that always comes to mind when comparing biology with physics, repeating time and again that the former does not seem ripe for the much more advanced objectives of the latter. Moreover, in the eyes of a physicist (the so-called "physicalist" position, often adopted in these comparisons), these terms of theory and concept are sometimes used a little strangely by biologists. Consider, for example, the expression used by R. Fisher for his "fundamental theorem of natural selection", which he saw as the equivalent in importance of the 2nd principle of thermodynamics in physics (on the concept of entropy), a very excessive statement, generated by enthusiasm more than by reality.

A very simple example can be noted in this conclusion, concerning what is called the logistic theory of growth. We have seen that it is based on a notion of distance measuring the instantaneous state of the variable with respect to its initial state and its final state. This amounts to *a priori* posing the existence of a stable stationary state at the end of the process, which is supposed to be a predetermined state. This can then be seen as genetic in nature, the constancy of parameters meaning native characteristics, neglecting the intervention of a fluctuating environment, i.e. a possibility of adaptation. Developing this "pre-training" point of view (let us say "instructionist") implies above all specifying this notion of distance (its metric *lato sensu*), leading to different forms of logistics. Of course, we can use other axiomatics, positing, for example, with L. von Bertalanffy, that growth activity results directly from the interplay of stimulation processes (anabolism) and inhibition processes (catabolism), the adult or saturated state being a simple consequence of this. In this simple example, we see that a reflection, combined with numerical experimentation on the assumptions underlying a type of model or on the value of parameters, can be a kind of prerequisite for the subsequent development of something that may have achieved the rank of theory.

Although this question is of general scope and is reflected in the choice of parameters for many biological process models, it is appropriate to focus our conclusion on the most essential points involved in the relationships that the biologist is led to establish with mathematical formalism. However, these biology–mathematics relationships go beyond the choice of the type of mathematical tools to be used, because any method carries an important biological and epistemological connotation that must be highlighted. By this we mean what is implied by the deliberate choice to be made in two kinds of dilemmas: continuous/discrete, on the one hand, and determinist/random, on the other. In addition, there is the question of the choice of the *observation scale*, reminding us of the statement by the physicist Guye that "it is the scale of observation that creates the phenomenon".

Discrete versus continuous

To shed light on the debate on the discrete/continuous dilemma, as it often arises in biology, let us refer to these two paths defined, on the one hand, by the Automata theory developed in the wake of J. von Neumann, and on the other hand, by the theory of differential dynamic systems derived from the thoughts of H. Poincaré. In the first case, we have seen that we are interested in the behavior of discrete units (such as cells) whose evolution is determined by their own instantaneous state and by the inputs from neighboring units, state and inputs being expressed in discrete language. To work like this is to establish *a principle of cellularity*. In the second case, which we have also illustrated under different situations, we put forward a completely different principle, according to which any process is the result of progressive transitions taking place on a certain substrate and therefore can be understood by continuous variables (possibly with discontinuities). It is a way to generalize the emblematic case of chemical kinetics, moving from the notion of molar concentration to that of population density, a completely classical approach since A.J. Lotka and V. Volterra. The problem is posed in terms of the spatiotemporal dynamics of continuous state transitions, which is finer and more flexible than reducing an evolution to a series of discrete jumps.

Of course, these two qualifiers of discrete and continuous are always more or less mixed. Thus, in any morphogenesis, any growth phenomenon is first based on a discrete 0/1 generation process resulting from mitotic activity, such as lateral budding in a mycelial or algal filament that will cause branching. Thereafter, for each cell of the branch thus initiated, it will be a question of a dimensional growth, i.e. an evolution in the continuous.

While these two approaches can be considered as complementary because they simply emanate from two different points of view, the fact remains that they are two *irreducible* paths in their very principle and in the assumptions they entail. Among the various debates thus generated, let us choose, to clarify our conclusion, the opposition between the discrete nature of A. Lindenmayer's automata (mainly aim morphogenetic) and the continuum of differential equations dealing with metabolic reactions generating a spatiotemporal organization (both physiological and morphogenetic) according to the classical analysis made by B.C. Goodwin.

For A. Lindenmayer and his L-systems, it is all about the behavior of each cell subject to precise transformation rules based on functions (current state/inputs) → (new state/outputs). On the other hand, B.C. Goodwin is a proponent of the notion of a *field* carrying gradients of activity whose evolution leads to singularities and eventually to the emergence of local structures. This leads him to the strong idea that morphogenesis is an inherently *robust* process[1]. If local structures are discrete in nature, they are very different from the changes of state exhibited by automata, because they are linked, not to *a priori* rules, but to the establishment of stationary asymmetries. More precisely, B.C. Goodwin takes up the idea of the genetic operon of F. Jacob and J. Monod and treats it mathematically, in the form of a network that describes the simultaneous and cybernetically coupled variations of different metabolites. Of course, it is necessary to add that which is a "mixed" approach, qualified as qualitative in relation to the continuous characteristic of differential equations, known as *"logical kinetics"* (Boolean), developed by R. Thomas, which we have also presented. If the current development of formalizations by networks (metabolic, genetic) shows the possibility of combining the two approaches, it is

---

[1] Goodwin, B.C, Kauffman, S., Murray, J.D. (1993). Is morphogenesis an intrinsically robust process? *J. theor. Biol.* 163(1), 135–144.

important to include in our general conclusion this discrete/continuous debate, because it reveals an important divergence from an epistemological point of view.

Thus, B.C. Goodwin's position highlights the concept of *regulation*, a fundamental notion for a physiologist that is of a variational nature, and therefore related to a theme that is absolutely essential in mathematical analysis. On the contrary, A. Lindenmayer offers us a view that could be described as both behavioral and instructional. This is reminiscent of a statement by philosopher H. Bergson abruptly saying: "There are no things; there are only actions" (in *Creative Evolution*). Every state is an instant view; what matters is, following Heraclitus, what flows, i.e. what the differential equations describing a given process express in transitions, shall we say.

In the end, this opposition can be summarized as follows: the Automata theory would only allow a purely formal modeling that would be analogical in nature. The cell would function as a computer that calculates its own state at each iteration and modifies it according to programmed instructions (the syntax of the production rules of the formal grammar defining the L-system at stake) without correspondence either with the biochemical level where interactions and transformations are performed, or with the existence of mechanical or topological constraints. The DNA control program is seen as the counterpart of the production rules, the driving force behind the automata's functioning. Clearly, it is the opposition between a *physicalist* point of view and a *symbolic* point of view. The first (differential equations) aims at a certain explanation of the phenomenon, while the second (production rules) is equivalent to a representation. On the one hand, we have an operational formalism capable of simulating the phenomenology of a population of cells or modules. On the other hand, it is the implementation of regulations that occur continuously during a dynamic process. In other words, the structuralism claimed by B.C. Goodwin (without focusing on what this term sometimes evokes as a debate) refers to an organized set of autocatalytic processes required to maintain a characteristic homeostasis, which gives it an *ontological* value (hence the term "natural model"), which is not provided by the symbolism inherent in any definition of automata. To conclude this debate, we can recall the comparison that can be made between powerful computer models simulating plant architecture and the explanation expected from a model showing how the plant regulates the growth and branching of its different modules on the basis of its physiology and physical and mechanical constraints. In other words, in substance, simulating does not necessarily explain two different views of what mathematics can bring to biology.

Microscopic versus macroscopic levels

We will not repeat the question of stochastic mathematical models, such as those we have seen in experimental designs or in quantitative genetics, combining a set of deterministic parameters and a stochastic part corresponding to what the model does not consider, including the randomness of the measurements. The question we would like to examine now is of a different nature and more general scope, relating to the type of quantities that can describe a phenomenon. Two kinds of characters are indeed to be considered according to the scale of perception of the phenomenon. We know that this is a classic question in physics where, for example, the state of a gas can be described *macroscopically* by temperature and density (or pressure and volume), while *microscopically*, what is at stake is directly the speed and position of the molecules of this gas that are subjected to interactions (collisions). The problem that statistical physics has been addressing since L. Boltzmann and W. Gibbs is to derive the overall properties of a system from the microscopic elements that make it up. It is the relationship between fine-scale statistical behavior and broader-scale deterministic behavior. Of course, biology also has these kinds of problems, which are well illustrated in population dynamics, for example. The overall refers to the average generation time and the average population growth rate; the microscopic is related to fertility and mortality rates, which are properties that arise at the individual level. It should also be added that, theoretically, this position results from what is called the "*ergodic hypothesis*" which, in essence, consists of saying that, at equilibrium, the average value of a microscopic (statistical) quantity is equal to the time average of this variable measured on a given particle at different times. For example, the average velocity of all particles at a given time (spatial aspect) is equivalent to the velocity of a given particle at different times (temporal aspect).

However, this is only a first distinction of scale. It should be pointed out that the macroscopic state of a dynamic system supposed to represent the evolution of a real material system results from the interplay of three kinds of variables, which physicists call external, mechanical and thermal. Let us review again the case of the kinetic theory of gases. For the first category, we have the volume and number of particles, and for the second category, the pressure and energy (which are averages of microscopic quantities). As for thermal quantities, they do not have a direct microscopic interpretation, being only statistically apprehended, such as the concept of entropy. The function called entropy refers to the number of microscopic states of the same energy level.

It is clear that this question of principle also arises in biology. Thus, in population dynamics, we often limit ourselves to two types of classical quantities, namely the size of each species and its growth rate (intrinsic growth + interactions). An approach inspired by statistical mechanics involves working on a population structured into age groups (according to the usual criteria). The number of such species is a mechanical quantity (resulting from the average of individual behaviors), while thermal quantities relate to other considerations, average generation time and entropy. The entropy of a population depends on the distribution of fertility and mortality rates (it is zero if there is only one fertile age class). Fertility and mortality are seen as analogues of an energy level. Its interest is to be a measure of the *complexity* of the population in terms of life cycles. Hence, it can be pointed out that A.J. Lotka's classic notion of stable population is equivalent to that of stochastic equilibrium in thermodynamic terms (invariance of age-class proportions).

Thus, such a probabilistic-based study of dynamic systems representative of a biological process leads to the introduction and quantification of the intuitive notion of organization (or disorder, associated with the concept of entropy) whose general significance in biology is known, from the morphogenesis of an organism during its ontogenesis to the balance of multi-specific associations or the kinetics of multistationary enzymatic systems. To shed some light on this overview, the reader may refer to the demographic treatment of the Leslie matrices we have discussed, consisting of assigning to each age group its size and rates of change (fertility, mortality in the sense of class transition). Hence a representation in the form of graphs expressing the genealogy of each individual, of which we can note in passing the difference with the families or cell lines of the L-systems.

These two absolutely classic continuous/discrete and determinist/probabilistic dilemmas that we have just summarized retain their importance, both epistemological and operational, as they determine the reasoned choice of the type of approach, i.e. the way in which a phenomenon is questioned and studied. However, leaving aside this kind of debate, which is not new, our conclusion must now be to highlight what the still recent evolution of biology implies.

The new trends in biology

Like the current academic designations for the name of major thematic streams in biology, a word should be said on the use of the terms "integrative biology" and "*systemic biology*" (or "*systems biology*"). In reality, these two qualifiers are by no means equivalent. The integration of two subsystems into a coherent whole means a reduction in the number of degrees of freedom with few or no innovative properties. On the contrary, the constitution of a complex system, with an increase in the number of degrees of freedom, is manifested by the appearance or emergence of *new collective properties*. These concepts concern a number of situations or processes such as, for example, a protein network, where complexity may simply result from competition between ligands for the same binding site, or from a change in molecular conformation, an unprecedented consequence of the joint consideration of several components or subsystems. In fact, an appropriate vocabulary would be to speak, according to the current trend, of "*complex systems*" to specify that there are emerging new properties in relation to the subsystems that constitute it. Does complex mean that what is not apparent results from the simple additivity of the parties?

Simple examples of networks were given to illustrate this systemic approach, showing the link between the principle of representation by oriented graphs and two types of formalization of biological regulations. The first example (Figure 3.12) illustrates the transition from classical differential formalism, such as that used by B.C. Goodwin (temporal organization of cellular activity), to its translation into a network whose nodes are genetic and/or metabolic in nature. The second example (Figure 3.13) takes up this idea but based on R. Thomas' qualitative representation of logical kinetics.

We know that the evolution of ideas in biology has led to the recent clarification and extension of the scope of this notion of complexity. Of course, it should not be forgotten that the objective was often of a practical nature: to be able to study a large and diversified set of data, i.e. to overcome what at first sight appears to be something very "complicated" without always clearly distinguishing it from what is called "complex". In fact, the term "complex systems" has a precise meaning, probably still in the process of being clarified.

This reaction is somewhat related to the use of statistical methods for multidimensional data analysis, which are supposed to overcome the "data overflow" (as Big Data now calls it) carried by large files (such as those that must be processed in ecology). However, we are well aware that the principle of any factorial analysis, which makes it original, is to identify what is called a "latent structure" or intrinsic organization that is hidden from the simple gaze of the experimenter. Although of different thinking, systemic innovation also leads to a new framework, that of networks whose specificity is to link levels of a different nature, according to the so-called multi-scale approach, such as the various metabolic and genetic circuits, or the components of any ecosystem *in situ*.

Two characteristics should be highlighted in this regard. On the one hand, the principle of causality, which we have had the opportunity to discuss, is here to be taken up in a new form, which is now distributed among several effectors. On the other hand, a network is generally not a fixed configuration (it is in fact a fuzzy whole, as links between nodes are subject to fluctuation), which is one of the new aspects of this general property of adaptation of any living being. The maintenance of what makes the *sui generis* nature of a living system is, as we know, its autonomy and homeostasis. Posing such a phenomenon in terms of a network connecting different entities at different organizational levels appears today as a new paradigm of biology which, without ignoring what other approaches have made possible, has become necessary for many phenomena.

The need for the connection between biology and mathematics is now recognized, regardless of its inherent or temporary limitations, for certain types of problems. There are indeed many concrete situations where the so-called naturalistic mind, not inclined to any mathematical formalization, retains a prominent place. This evidence leads us to recall B. Pascal's famous distinction between the spirit of finesse and the spirit of geometry. The spirit of finesse is based, he says, on principles that are "in common use" and for which "it is only a question of having good eyesight". As for the spirit of geometry, B. Pascal clearly considers that "geometricians have a straight mind, but only if all things are well explained to them by definitions and principles". Nowadays, this reflection can be interpreted as the distinction between *explanation* (mind of geometry) and *understanding* (mind of finesse).

As we have often pointed out, the particularity of mathematics is well in this spirit of geometry: it is always based on definitions and hypotheses (even if it means restricting the meaning of nuances so strongly in biology, where the importance of variability often leaves doubt on the relevance of too-strict typologies[2]), resulting in a well-defined field and way of investigation. Its natural outcome towards an explanation (via modeling) is necessarily marked by this limitation of its field of study and the point of view adopted. However, this is only a trivial observation for any methodology that can only provide its own explanatory potential.

But what exactly is meant by this term "*explanation*", which is opposed to simple "*description*"? If we are to limit the debate, let us refer to an interesting classic definition, which says that "a fact has been explained when we have a mathematical or logical form capable of generating the description of that fact" (J. Largeault). In connection with this property of being able to generate a structure or function, we can consider, with R. Thom, that the causal explanation thus posed allows a reduction in the arbitrariness of the description, purifying in a way the diversity of descriptive characteristics, which are of very variable relative importance, and are sometimes more or less redundant. Nevertheless, there is no agreement on this proposal, as some believe that a simulation with a good prediction can be used as an explanation since it provides information on how the system

---

2 On this general theme of boundaries and categories, see Parrochia, D. (1991). *Mathématique et existence*. Éditions Champ Vallon, Seyssel.

works when seen as a black box. Consequently, from a dynamic point of view (an essential point since any system is subject to constraints that lead it to adapt and evolve), *the explanation must be consubstantial with the functioning*, and not only with the structure. All this cannot prevent us from thinking of the image of a projective geometry. If the model always has something to do with reality, it is, by its nature, equivalent to a *projection of the mind (the chosen formalism) on the reality*, of which we cannot say *a priori* the part of invariance that this projection respects and the part of deformation or masking that it cannot avoid.

Thus, we tend to bridge the gap or distance between the mathematically expressed biological object and the concrete real object, whereas in itself, the explicit mathematical structures can at best only be isomorphic to reality, or, to use L. Wittgenstein's words, mathematics is nothing more than grammar rules, distinct from the meaning they convey. While there is no question of rejecting the idea that the mathematized object remains in the domain of ideas without close bijection with the concrete real object, the question arises as to whether or not the mathematical formalization of life can be endowed with the status of prolegomena that would initiate a certain understanding of the studied phenomenon. Of course, the equationization of a process and its insertion into a network of relationships remains essentially a representation, the comparison of which comes to mind with R. Magritte's famous painting entitled *The Treachery of Images*. This painting represents a pipe whose title states: "this is not a pipe", meaning that its painting is not reality, but its representation by the eye and the brush. However, this risks being a prevarication, which prevents us from admitting that a set of properties highlighted by a mathematical approach is quite likely to contribute to better understanding, if not the essence, then at least the expression of the living, what makes it manifest itself to us and makes us recognize it as such. In other words, let us repeat, mathematics can be a tool of perception which, without being an end in itself, has no reason to exist but to lead towards a better knowledge of living things as they appear to the observer who measures and formalizes them, and then, eventually, simulates, predicts and reconstructs them.

The meeting of these two disciplines cannot *ipso facto* lead to a duly argued demonstration. On the other hand, what is important to note is that the intrinsic properties of living that the mathematical tool can detect on a given system or organism constitute an original asset, not redundant with the data of experience. Within the framework of the mathematical concepts used, its value is to be provided with a certain coherence, allowing it to claim to improve knowledge of living. It is therefore to agree *a priori* with the postulate that an important mathematical property of a given biological system corresponds to an equally important property of a biological nature.

Underlying these discussions is the unavoidable anthropic principle, namely the recognition of an agreement between the order of nature and the requirements of rationality and coherence of the human mind. Many authors have expressed it, such as A. Einstein, whose famous aphorism is well known: "The most incomprehensible thing about the world is that it is understandable". In the absence of the further development that this subject deserves, and although our view is scientific and not philosophical, we cannot avoid this fundamental question, which is the existence of a double enigma, as formulated by A. Koyré, before it was taken up and specified by K. Popper. In short, let us say that today it is no longer a question of solving only the enigma of the physical universe, but also that of the human spirit that is attached to its study[3].

In the end, it is a question of assessing whether the use of such concepts implemented by a particular mathematical method to study a particular biological object makes it possible to *better identify* this object by a *sui generis* set of characteristic properties that are considered essential. Moreover, the search for these, like an identity card (containing generic characters and particular signs), has always been the primary scientific objective of the human mind when confronted with the enigma of nature, including the existence of the famous biological variability that has surprised many physicists, such as M. Delbrück, interested in the modeling of living. In this regard, we can recall the naturalistic position of the first field observers of the time when biology was called "natural history". For them, the first task was to recognize and specify (before classifying) the species, the organ, the developmental stage and even the environment that shapes it, in order to ensure that it was used wisely and without making unfortunate errors in diagnosis. The proper use of medicinal plants illustrates this old imperative, which goes beyond a simple static morphological description. A detailed study of the structure and, above all, the functioning of a given biological system also requires this basic principle of signage.

These are the challenges of this necessary encounter between biology and mathematics, an encounter that aims to better understand and bring coherence to what is being studied experimentally elsewhere. In other words, we set ourselves, if only implicitly, the ideal objective of giving meaning to what we are studying. By this expression, *"to give meaning"*, we mean to obtain a set of properties that seem to be specific to the object or process under study and thus contribute to clarifying its identity, a little like a detailed definition, with what is specific to it and what has a more general scope at a broader level, such as species or cell population. This objective of research into the intrinsic properties, distinct from (and possibly complementary to) phenomenological reconstruction, goes far beyond what is called a description, justifying, if necessary, the fundamental interest of a mathematization of living within a well-defined conceptual framework.

---

3 Cited by Parrochia, D. (1991). *Le Réel*. Bordas, Paris, 160 and 163 *sq*.

Glossary

This glossary contains various terms or names used in mathematics or biology, in order to place them in a wider context while specifying the more contextualized use that is made of them at various points in the text.

Bifurcation

In mathematics, this term designates the attribute of a dynamic system by which it shows an abrupt change of trajectory caused by a small variation of a parameter. This is then a modification of the system stability, going against the principle of structural stability, which is the objective (see entry for "stability").

The classic example is discrete logic: $y(t+1) = a\,y[1-y(t)]$, the dynamic behavior of which qualitatively varies in response to the value of parameter a. A progressive variation in a leads to a change from a single stable fixed point (asymptotically) to the establishment of an oscillating dynamic, which is itself variable in response to a sequence of doubled time intervals (number of oscillation peaks) before generating chaotic behavior. This is characterized by the unpredictability of dynamics and by its dependence on initial conditions.

The remarkable point is the uncertainty produced by a very simple autonomous model of this kind, which is *a priori* deterministic, since it does not contain any random terms. A well-documented classic type is the *Hopf bifurcation*, which relates, from a theoretical point of view, the bifurcation property to the evolution of the complex nature of the eigenvalues of the system (studied on the linearized system), in particular the change from behavior at a stable fixed point (damped oscillations) to the establishment of a stable limit cycle. The general issue of a qualitative change of dynamic is particularly relevant to multistationary systems.

The evolution of these systems is marked by the existence of attracting sets attached to each of the stationary points (stable or unstable). The possibility of moving from one region to another depends, on the one hand, on the initial conditions and, on the other hand, on a variation in the value of the parameters.

This bifurcation property, as an abrupt change in dynamic, can be observed in biology, where processing the question through mathematics can be a basis for its formalization. A good example is provided by the *sexualization of a caulinary meristem* in plants. At a given stage of ontogenesis and taking into account the appropriate environmental conditions (length of the day), this meristem undergoes a change of state that leads to a series of stages (flower induction, then flower evocation and finally initiation) that modify the nature of neoformed tissues. The meristem now no longer generates a sequence of plant metamers (internodes and leaves), but instead a set of elementary floral parts, which are themselves spatially organized in a sequence of flowers. A new morphology results from this, moving from phyllotaxis of vegetative organs to the typical architecture of reproductive organs (inflorescence). This transition in development from vegetative to reproductive is generally not simultaneous on different axes of the same plant.

Another remarkable case is the behavior of the ameba *Dictyostelium discoïdeum* as a function of the nutritional environment. This protist has two types of development, either a single-cell form of free mobile protists or, in the event of a nutritional deficiency and by chemotactism, their aggregation in characteristic compact colonies (pseudoplasmodia). This multi-cellular phase is marked by triggering the process of differentiation that leads to sporulation. In this case, the bifurcation is reversible, mainly determined by the environment, which leads to an alternation between the free form and the aggregate form.

Calculation of the variations or variational calculus

This designation is connected with the study of the optimization of a function (a given function, a function of a function, an integral, etc.). Seeking out a peak in fact corresponds to the optimum with regard to the practical objective that the given function expresses, hence the interest of employing it in multiple fields. For example, in differential geometry, there is the geodesic or shortest path that is sought between two points in a given space. In physics, these methods are at the basis of the Maupertuis principle of least action. It is also used in biology, for example, as an orientation strategy of ontogenesis.

This branch of analysis was developed on the basis of the Euler–Lagrange equations, then revised by the Hamilton formula. Its applications constitute what is known as "optimal control", using control variables.

Cambium

See "Meristem".

Caulinary

Relates to a plant stem or branch.

Conservative

In biology, this term describes any system that is not subject to a process of mortality. It concerns, for example, *in vitro* cell cultures in a reactor such as a chemostat with renewal of the environment allowing generation and biosynthesis, but excluding all mortality (this is a specific case of a reactor). In mathematics, a conservative system is characterized by the existence of a function f (known as the first integral of the system) that remains constant throughout any trajectory. In mechanics, the term applies to any force whose effect is independent of the pathway previously taken (lack of memory).

On the other hand, we have the behavior of systems with memory, such as *hysteresis* in which we experimentally observe that the variation of the control variable does not lead to the same dynamic for increasing or decreasing variation, for example, in physics, the variation of the intensity of a magnetic field on the behavior of a magnetic material. This type of behavior is observed in biology for various processes, in particular in biochemistry or in ecology. For example, the regeneration capacity of a prairie is a function of the variation, increasing or decreasing, of the density of herbivorous predators, which determines the number and stability of the stationary states.

A specific case of non-conservative process is that of the Markovian processes, discrete models in which any state depends only on the state immediately preceding it (which means it is qualified as a process without memory).

Continuous, discontinuous, discrete

For these commonly used terms, here we specify their interpretation that is used comparatively in biology and in mathematics.

Biologists are familiar with the term "discrete", which relates to any generation or any abscission of a specified element (cell, module or individual in a growing population), such as processes of birth and mortality for which proven mathematical models are available to them. However, many biologists add to this the idea of a principle known as "cellularity", which would be an unavoidable basic principle for a formalized approach to all morphogenesis. Constructed as a hypothesis, this principle actually opposes the idea of continuous.

In biology, "continuous" relates to any element (of matter or of dimension) of any object that cannot be specified and discretized macroscopically. The mathematical reference (concerning analysis and topology, outside the set theory) is the differential whose significance lies in the notion of a limit $\Delta x \rightarrow dx$ (see the change from the secant to the tangent in the geometrical interpretation of the derivative of a function). In the same way as the use of vector calculus in mechanics of continuous environments (e.g. in fluid mechanics), biology takes on an analogous position for local analysis of activity (growth, metabolism) within a given biological field. The remarkable thing is that the idea of a continuum is presumed in advance in biology. In contrast to the "cellularistic position", which favors the idea of discrete states for which we seek to establish vocabulary (over and above the simple fact of the neoformation of entities), the idea of continuity is equivalent to attributing fundamental importance *ipso facto* to the notion of *transition* (infinitesimal) of a given state towards a later state, depending on the connotation that chemical kinetics provides. There are many differential models that have a reaction interpretation of this type. This principle of elementary transformation per unit of matter and unit of time is based on two fundamental ideas; movement or flow, and form (*lato sensu*).

The distinction between discrete and continuous is not reduced to a simple question of scale of perception, but instead refers to the importance to be given, or not, to a double notion of transformation, whether it is in a metabolic or dimensional sense of the term, and that of singularity in a functional sense within a continuum. In addition, what is continuous presents the advantage of being associated locally with the discontinuous, meaning with the existence of a point of rupture in the evolution of a system. This is the notion of bifurcation, an essential marker of certain dynamics, as important in biology as in mathematics (see entry for "bifurcation").

Finally, let us note that a kind of bridge can be established between the strictly discretized design of a series of different states and the design of a continuum of elements of matter that are subject to transformation. This is the approach known as "kinetic logic" that brings together the Boolean logic (states 0/1) and the differential formalism, with the objective of formalizing the fundamental notion of regulation. More generally, the formalization in networks associates the two points of view of discrete (nodes) and continuous (transitions by directed graphs).

Growth

There is a great variation in biological growth depending on the relative part of the basic processes, division and increase in size (elongation) of cells. Many mathematical models ("growth laws") express the process phenomenologically and several of them have biological bases for interpretation (structured models).

Various types of growth need to be distinguished depending on the typology of the process, in particular for plants, where it is necessary to distinguish between "polarized growth" (e.g. in filament organisms, fungi or algae, where the development of the filament is determined by the mitotic activity of the apical cell) and "distributed growth", where the activity is spread over a field of growth (such as in the surface extension of a leaf blade). From a cellular point of view, the extension of the wall is the result of several elementary mechanisms: hydric flows that modify the turgescence, extensibility/relaxation of the wall and enzymatic incorporation of new metabolites (microfibrils of cellulose assembled by a matrix of hemicelluloses and by pectins). In addition, in contrast to animals, plants are characterized by symplastic growth, a term that indicates the maintenance of liaisons between cells. By the term "symplast", we understand the continuum between adjacent cells, a continuum that is ensured by the existence of channels (plasmodesmata) passing through
the walls.

The old, classic definition of growth as an irreversible increase of the dimensions of an organ must be reconsidered. On the one hand, the phenomenon of growth can be periodic in nature, sometimes with clear phases of latency or dormancy. On the other hand, following the example of demography, we cannot separate out various morphogenetic processes that coincide in a coordinated manner in the dynamics of one and the same ensemble that is developing, for example, the generation of new elements, their increase, their differentiation and then their aging and mortality. The term growth must therefore be taken *lato sensu*, since it pertains to a unit ensemble of increases, decreases and changes of state that interlock within a given biological system.

Heteroblasty

Heteroblasty is a property of ontogenesis that consists of an unequal distribution of a quantity or of a biological activity depending on the position within the organism. This effect of position, added to the implementation of axes of polarity, characterizes in particular the set of the processes of embryogenesis, whether animal or plant. This property is also highly important in the post-embryonic development of a plant axis.

For example, the dimensions and form of leaves vary as a function of the insertion level on the stem, which contributes to the construction of a characteristic architecture and to the photosynthesis phenomenon of the plant (capturing light energy). A heteroblastic function allows these variations to be mathematically expressed, such as the evolution of the growth function parameters, whether for leaves or internodes, depending on the position on the carrying axis.

Hypothetico-deductive method

This is a classic experimental approach that consists of formulating a hypothesis, observing the consequences that allow conclusions to be drawn by logical deduction about the validity of this hypothesis. This position of principle, traced back to R. Bacon (1267), who noted the advantage presented by mathematics for the study of natural sciences, was justifiably highlighted by C. Bernard (1865), insisting on the essential role of the hypothesis: "Without a hypothesis, meaning without an anticipation in the mind of the facts, there is no science"[1]. This is the scheme of work: "The fact suggests the idea, the idea directs the experience, the experience judges the idea". This refusal of precedence for "raw facts" is picked up by K. Popper in opposition to the empiricism of the inductive approach.

Latent structure/latent variables

This is a founding notion of factorial analyses expressed by the following fundamental hypothesis: the correlations between observed variables result (in the same way as a phenotype) from their dependence with respect to hypothetical variables known as latent variables. Thus, we distinguish between "observed variables" (or manifest variables) and "latent or underlying variables" (also called factors or components). The term latent is used here to mean something hidden (or virtual), entirely different from the meaning in the designation of the latency phase that is used in process kinetics (prior to the beginning of strong exponential growth). These latent variables are not physically observable, but they are brought to light by the calculation (see the method of principal components analysis in linear algebra).

---

[1] Bernard, C. (1974). *Principes de médecine expérimentale*. PUF, Paris, 77.

This hypothesis is linked with a principle of *conditional independence*: the observed variables are conditionally independent of the latent variables. Let us give a reminder of what conditional probability is: for example, in the case of three observed variables X, Y and Z, we say that X is independent of Y under the condition Z if:

$$P(X,Y|Z) = P(X|Z).P(Y|Z)$$

We refer back to Fisher's elementary statistical notion of *"partial correlation"* $r_{XY.Z}$, which is the Bravais–Pearson correlation between X and Y when we remove their dependence on Z.

The reason for the existence of this notion of latent structure is to provide an explanation of the diversity of the observed correlations within a set of data that pertain to a given phenomenon. It is of interest in the interpretation (in biological terms) of the nature of latent variables that are hypothesized in this way. Its explanation is found in moving from what is apparent to what is underlying.

The reference to this principle (often not explicitly explained) attributes a specific status to the factor analysis, which makes it stand out from procedures of descriptive statistics (ordered in a set of observables) or inferential statistics (tests on samples).

Meristem

This is a botanical term that describes a localized ensemble of young totipotent cells, in a state of multiplication prior to any differentiation. There are two types of meristems depending on their location and the nature of their descendants. By "primary apical meristem", we understand the region of a more or less ovoid shape located at the extremity or in the axil of an axis (bud of a stem, tip of a root) where new cells are generated by directed mitotic activity. These well-defined territories, called "vegetative points", are at the origin on a caulinary axis (stem) of new internodes (extension of the stem) and new leaf primordiums (see "Phyllotaxis"). The cambiums or secondary meristems are differently organized in the form of cellular foundations, which are more or less continuous, located inside a plant axis (their ring shape corresponds to one of the cases envisaged by A. Turing in his reaction–diffusion systems).

Their mitotic activity is at the origin of two types of tissues depending on the direction, for example, the neoformation of conducting tissues, liber (or phloem) on the external side and sapwood or young wood (xylem) on the internal side. The formation of cork (suber) is also of this type, resulting from the activity of a layer known as "subero-phellodermic", like the libero-ligneous cambium, that participates in the radial growth of plants. The dynamic of the cambial activity is particularly important in arborescent species whose growth is specifically marked by an ensemble of mechanical, radial and longitudinal constraints.

Metamer

A metamer is a morphological unit or constitutive module of certain developing organisms along a polarized axis. Metamerization or segmentation is a typical plan of a biological organization. In animals, each segment can be repetitive (e.g. in certain worms or annelids) or, on the contrary, can be the site of a particular morphogenesis (e.g. in insects).

In the higher plant, this term, also known as "phytomere", designates all the organs generated by an apical caulinary meristem (internode, leaf, axial bud). Outlined between two consecutive nodes, this module is generated in each of the meristematic operation cycles. Very generally, these successive modules, although they are repetitive in broad terms, present characteristics of growth (dimensions) and of morphogenesis (ramification, sexualization) that are variables depending on their position on the axis. This property, said to be "heteroblasty", is an important characteristic of plant development (aerial organs) that intervenes in the architecture and physiology of plants.

Mitosis

We distinguish two types of cellular divisions depending on their direction. A cellular division is said to be "anticlinal", or respectively "periclinal", if the axis of division is perpendicular, or respectively parallel, to the surface of the tissue. These distinctions are of particular importance for plant tissues, conditioning the variations in shape of an organ. See their correspondence with the notion of main axes of growth of an apical meristematic dome in 3D.

The two daughter cells of a mitosis need to be considered *a priori* as possibly dissimilar. The occurrence of asymmetric mitoses, observed in highly diverse organisms, contributes to the diversity of cellular populations. For example, it is an important morphogenetic property in filamentous organizations of apical growth (fungi, algae). The daughter cell in apical position takes over from the mother cell (growth without differentiation), whereas the sub-apical (of different sub-dimension and polarity) can be at the origin of the branching of a side shoot.

By "meiosis", we understand a division of a particular type, which intervenes during the formation of gametes for which the haploid number (n chromosomes) requires a chromatic reduction of initial somatic cells to $2n$ chromosomes.

Normal probability law

This law, also known as the Gaussian or Laplace–Gauss law, describes the probability distribution of a continuous random variable, of parameters μ (expectation) and σ^2 (variance) and of the field of definition $[-\infty, +\infty]$. By a transformation, we obtain the "normal centered reduced law" for parameters 0 and 1. Graphically, the distribution law (sum of probabilities) is an asymptotic, sigmoid and symmetrical curve. The law of density (probability by value) gives a bell-shaped symmetrical curve (one maximum in μ and two inflexion points in $\mu \pm \sigma$).

The normal law is stable by additivity (the sum of independent Gaussian variables is also Gaussian), hence its statistical interpretation as a law of measurement errors. The normal law is also stable by linear transformation.

This law generalizes into a multi-dimensional normal law.

Null hypothesis/alternative hypothesis

These are designations used in all statistical inference. Consider the simple example of comparison of the means of such a variable (like a dimension or content) measured for two samples, let us say $m1$ and $m2$. That is, $\mu 1$ and $\mu 2$ are the theoretical values that the corresponding parent populations would give us if these were accessible to us (in precise terms, mathematical expectation). The comparison test of the two averages is based on the hypothesis, known as the "null hypothesis" H0, of an absence of a difference in the level of parent populations: $\mu 1 = \mu 2$. It is not a case of mathematically demonstrating an equality but instead of being able to attribute a probability of occurrence to the difference empirically observed between the samples, meaning rejecting the null hypothesis in favor of a complementary hypothesis, known as the "alternative hypothesis" H1. The conclusion does not express a strict equality, but instead the existence of a significant difference. Depending on the context, H1 corresponds either to a simple inequality $\mu 1 \neq \mu 2$ or to a relative order $\mu 1 < \mu 2$ (or vice versa). The rejection decision is probabilistic in nature (see "Risk of errors").

Ontogenesis

This is the ordered sequence of morphological and physiological processes that are achieved during the development of a living organism or of one of its parts or sub-systems. Various characteristic phases thus follow, first during the embryonic life right from the first cellular proliferations and the differentiation of the first outlines, then during postnatal life in which morphogenesis, growth and aging are embedded, spatially and temporally.

Plant ontogenesis is distinguished from animal development by a continuous embryogenic property. Any higher plant (or any filamentous organism) is the almost permanent site of morphogenetic processes, since the generation of new organs (through maintenance of totipotent meristems) is combined with abscission (natural pruning) of organs that have become inactive. The life of a plant is confused with its growth: the plant is by nature an evolutive metapopulation of organs. The development of large organisms (arborescent species) leads to characteristic architectures (morphological notion that is more than a simple outline or a silhouette). Formation and extension of these plant architectures can be done by reiteration or taking up (copying) modules or sub-systems that were previously differentiated. Similar questions relate to the development of fixed colonial organisms such as corals. Their ontogenesis is presented as a spatial assembly of individual entities (polypes), and as their morphophysiological integration in a complex structure known as "superorganism".

Optimal control (or command)

Methodology arising from Maupertuis's physical principle of "least action" (or minimal work), which is widely discussed in mechanics and optics, intended to control how a process plays out via one or several exogenous variables known as "control variables". It was developed in research work based on the Pontryagin school of thought, setting up the necessary/sufficient conditions for a solution. Prior to this, mention must be made of the property of multistationarity of many dynamic systems, where their evolution towards such or such a fixed point depends on the value of certain parameters. Mathematically studying how these, depending on their interval of variation, can determine a given development strategy is equivalent to considering them to be control variables, hence a possibility of command of the process.

From a theoretical point of view, the principle of least action has a finalistic connotation depending on the phrase that can be attributed to P.L.M. Maupertuis: "If a change occurs in nature, the quantity of action required to accomplish it must be the least possible".

In addition to its multiple applications in physics and mechanics, the idea of optimal control has great practical importance in biology today, for example, for the choice of a development strategy of a cultivated plant (encouraging production of biomass in the form of vegetative organs or seeds) or the optimization of a bio-industry process (cellular cultures *in vitro* in reactors).

Plastochron

This is a botanical term that designates the time lapse separating the differentiation of two consecutive metamers (internode + leaf) of a stem or of a branch of the higher plant. The plastochron *stricto sensu* corresponds to the period of cyclic operation of an apical caulinary meristem. The volume of the apical meristematic dome varies, increasing by mitotic activity during generation of a new leaf initium and becoming minimal after uplifting or emergence of the constitutive draft (leaf primordium). In practice, the term is used, under the designation "phyllochron", to monitor the kinetics of macroscopic appearance of successive leaves on an axis, where the estimate can be made in a non-destructive manner (without dissection of the bud), by basing itself on the acquisition time of a dimension given *a priori*. The plastochron must be considered, on an agreed basis for its estimation, as a variable that is likely to fluctuate during caulinary ontogenesis.

Potential function

Notion relating to a dynamic conservative system, known as a "gradient system", in which the displacement of any point in the field in question (e.g. an element of matter in a growth field) only depends on their position (and not on the path taken). The elementary circulation of a vector V is V(x)dx. This property is associated with the local value of the gradient of a given function, known as a potential function φ. The speed of displacement to the position x_i is expressed by: $v_i = \partial \varphi / \partial x_i = grad\, \varphi(x)$. Its representation on a graph sets out the existence of local extremum and barriers to potential.

Refutation of a theory

The hypothetico-deductive method leads to being able to accept, or refute, the hypotheses laid down *a priori*. K. Popper insists on the importance of the refutation operation (*The Logic of Scientific Discovery*). It is a case of refutation because the hypotheses are shown to be false.

According to K. Popper, any scientific approach (except for mathematics) must, instead of demonstrating that a proposal is true, establish that it cannot be disproved. Actually, the value of a theory thus rests on its current resistance to attempts at refutation, rather than on obtaining proof of truth.

Risk of error

Any decision about statistical inference is a choice between acceptance and rejection of the null hypothesis H0. This approach is probabilistic in nature since any test statistic S is a random variable. In any decision-making, two different situations can appear. We can, in effect, either accept a false hypothesis (risk of first-type error α) or reject a true hypothesis (risk of second-type error β), which is written in terms of probability as: $P(H_1|H_0) = \alpha$; $P(H_0|H_1) = \beta$.

The complements of these risks are, respectively, named the "confidence" and "power" of the test. In practice, the risk of the first-type error α is chosen *a priori*, hence its arbitrary character, for example, the value $\alpha = 0.05$ that is often adopted, leading to the expression "having a chance of 5 in 100 of making a mistake". The logic situation is given in the table of occurrences with two inputs, true hypothesis/selected hypothesis:

Probability of the decision	H_0 true	H_1 true
Accept H_0	$1-\alpha$	β
Reject H_0	α	$1-\beta$

Stability

Stability is a property of a stationary state of a dynamic system (cancelation of state variables $dy_j/dt = 0$), characterized by the fact that following any deviation or disturbance away from the stationary state, the system returns to its stationary position. Usually this is an asymptotic stability known as "Lyapunov stability". Stability comes from the properties of the Jacobian J (determinant of the matrix of partial derivatives of the linearized dynamic system around the stationary state). It is conditioned by the existence of negative eigenvalues (in their real part) of J. See the text for various kinds of stability, punctual (fixed point) or oscillatory, with the remarkable case of the stable limit cycle.

A multistationary system can present one or several metastable states. Such states correspond to the existence of local energy minimums (wells) associated with barriers of potential.

The designation "structural stability" describes the resistance of the behavior of a system, meaning that any variation in the value of its parameters does not qualitatively modify its dynamic.

Stationary state/equilibrium state

By "stationary state", we understand the state of a variable x that does not vary over time: $dx(t)/dt = 0$. This term generally corresponds to the term *steady state*. However, depending on the context, there can be some ambiguity with the constant nature of a speed of transformation, whether absolute $dx(t)/dt = Cte$ or relative $(1/x)dx(t)/dt = Cte$. For example, *steady state* sometimes describes a phase of exponential growth (also designated logarithmic phase) that is defined by the invariance of the growth rate (and not of the variable itself).

"Stationary state" and "equilibrium state" are generally seen as equivalent. The reference to equilibrium means more specifically the equality of input and output flows of a given system (e.g. a reversible chemical reaction at the equilibrium state). We also talk about "thermodynamic equilibrium" for the state of a system with constant entropy: $\Delta S = 0$, and, *a contrario*, of "systems far from equilibrium" or "dissipative systems".

The term "multistationarity" designates the property of a dynamic system to include several stationary states of variable stability.

Statistical inference

This term designates the statistical tests approach, for example, in the comparison of samples, with a view to having a precise idea of the homogeneity of a population or of the efficiency of an experimental treatment. "To infer" is to move from a given level (measured sample) to a more general level, which is that of the inaccessible theoretical population (parent population) for which the sample is a representation of a probabilistic nature (i.e. randomized type). The estimate of the law of probability of a variable and of its parameters from sample data also arises from this approach. To infer means to interpret in order to conclude.

All inference, since it is probabilistic in nature, is attributed with a certain *risk of error* (refer to this term). From a philosophical point of view, to infer is to advance from assumptions to the conclusions that they potentially carry.

Any inference work uses a variable known as a "test statistic". Classic inference, known as "parametric", is based on the probabilistic properties of this test statistic (for which tables provide values for a series of probabilities). Using it often requires, for example, the variable studied to be normal (Laplace–Gauss law). In the non-parametric tests, the approach is different, mainly based not just on the statistical distribution of values, but on the information given by their classification in a ranking. These non-parametric tests are generally more robust (more flexible application conditions) than parametric tests, where these are more powerful in terms of risk of errors of second species.

Vector analysis

The aim of vector analysis is to study the spatial distribution of a given variable. More specifically, there is a focus on local properties based on vectors associated with all material elements in the field in question, for example: field of speeds of growth in an appropriate Euclidian space to represent a given biological object, such as $\mathbb{R}^1$ for the extension of a root, $\mathbb{R}^2$ for the extension of a flat leaf blade (unfallen leaves) or $\mathbb{R}^3$ for mitotic activity in various directions (anticlinal or periclinal in particular) of an apical caulinary meristem.

Classic first-order operators (first derivatives) are the gradient, divergence and rotation. They are expressed as a function of a basic operator denoted *nabla*, written ∇, which expresses the gradient of the quantity in question (partial derivatives with respect to the reference directions). In practice, by "gradient operator", we more specifically understand the variation of a scalar, such as a concentration. Expressed as the value of a function F, this gradient is written as:

$$grad\ F \text{ or } \nabla F = (\partial F / \partial x,\ \partial F / \partial y,\ \partial F / \partial z)$$

A gradient has the nature of a vector.

The "divergence" operator describes the variation of the components of a vector $\vec{V}: div\vec{V}$ or $\nabla \vec{V} = (\partial V_x / \partial x, \partial V_y / \partial y, \partial V_z / \partial z)$.

Finally, the variation of the direction of $\vec{V}$ is given by the *rotational* operator, which leads to intervention of differences of the components of the divergence vector two by two. This is the vector $rot\vec{V} = \nabla \wedge \vec{V}$ ($\wedge$: vector product bringing the angle between the directions in question into play).

By way of respective examples of the use of these operators, we have the evaluation of local variations in concentration of a given substance (importance in diffusion or convection processes), in the speed of growth (regionalization of an inhomogeneous field) and in the anisotropy of growth (vorticity or change of direction of growth).

By way of a classic second-order operator (second derivatives) on a scalar or on a vector, we have the Laplacian, written as $\Delta\ or\ \nabla^2 : \Delta F = \nabla(div F) = \left(\partial^2 F / \partial x^2, \partial^2 F / \partial y^2 F / \partial z^2\right)$. Expressing variations of a speed, it takes on the nature of an acceleration, involving, for example, the speed of evolution of a physical quantity in a given field.

Vector analysis is extended with *tensor calculus*, which considers the partial derivatives of the components of the speed vector in such a way as to express all the constraints that any element in the field is subjected to. A tensor is an extension of the notion of vector. It takes the form of a symmetrical matrix whose diagonal elements are the normal constraints and whose extra diagonals are the tangential constraints, for example: growth or deformation tensor (*strain rate tensor*), vorticity tensor and conductivity tensor (study of flows of matter such as hydric flows in plants).

References

Below are some general references, supplementing those given throughout the book on more specific topics.

Arino, J. (2001). Modélisation structurée de la croissance du phytoplancton en chemostat. PhD thesis, Université Grenoble 1, Grenoble.

Atlan, H. (1992). *L'Organisation biologique et la théorie de l'information.* Hermann, Paris.

Auger, P., Lett, C., Poggiale, J.C. (2010). *Modélisation mathématique en écologie.* Dunod, Paris.

Bailly, F., Longo, G. (2006). *Mathématiques et sciences de la nature. La singularité physique du vivant.* Hermann, Paris.

Bertrandias, J.P., Bertrandias, F. (1997). *Mathématiques pour les sciences de la vie, de la nature et de la santé.* EDP Sciences, Les Ulis.

Bourgine, P., Lesne, A. (eds) (2006). *Morphogénèse. L'origine des formes.* Belin, Paris.

Buffon G.-L. (2017). *Œuvres complètes de Buffon.* HardPress, Miami, USA.

Buis, R., Lück, J. (2006). Comptes rendus biologies. *Académie des Sciences de Paris*, 329, 880–891.

Buis, R. (2016). *Biomathématiques de la croissance. Le cas des végétaux.* EDP Sciences, Les Ulis.

Caswell, H. (1989). *Matrix Population Models.* Sinauer Associates Publishers, Sunderland.

Chauvet, G. (1987). *Traité de physiologie théorique* (3 volumes). Masson, Paris.

Cherruault, Y. (1983). *Biomathématiques.* PUF, Paris.

Cherruault, Y. (1998). *Modèles et méthodes mathématiques pour les sciences du vivant.* PUF, Paris.

Demetrius, L. (1979). Démographie, taux de croissance et entropie. *Population*, 4/5, 869–882.

Demetrius, L. (1983). Statistical mechanics and population biology. *Journal of Statistical Physics*, 30, 709–753.

Douady, S., Couder, Y. (1993). La physique des spirales végétales. *La Recherche*, 250, 26–35.

Douady, S., Couder, Y. (1996). Phyllotaxis as a dynamical self organizing process part I: The spiral modes resulting from time-periodic iterations. *Journal of Theoretical Biology*, 178, 255–274.

Françoise, J.P. (2005). *Oscillations en biologie*. Springer, New York.

Gayon, J., Petit, V. (2019). *Knowledge of Life Today*. ISTE Ltd, London and John Wiley & Sons, New York.

Glade, N., Stephanou, A. (eds) (2013). *Le Vivant discret et continu*. Éditions matériologiques, Paris.

Goldbeter, A. (1990). *Rythmes et chaos dans les systèmes biochimiques et cellulaires*. Masson, Paris.

Goodwin, B.C. (1963). *Temporal Organization in Cells*. Academic Press, London.

Günther, B., Morgado, E., Jiménez, R. (2003). Homeostasis and heterostasis: From invariant to dimensionless numbers. *Biological Research*. 36(2), 211–221.

Istas, J. (2000). *Introduction aux modélisations mathématiques pour les sciences du vivant*. Springer, New York.

Kauffman, S. (1993). *The Origins of Order: Self-organization and Selection in Evolution*. Oxford University Press, Oxford.

Kauffman, S. (1995). *Antichaos et adaptation*. Pour la Science, Paris.

Keller, E.F. (1999). Le Rôle des métaphores dans les progrès de la biologie. Institut Synthélabo, Paris.

Kostitzin, V.A. (1937). *Biologie mathématique*. Armand Colin, Paris.

Lesne, A. (2008). Robustness: Confronting lessons from physics and biology. *Biological Reviews*. 83, 509–532.

Lesne, A. (2009). Biologie des systèmes : L'organisation multièchelle des systèmes vivants. *Méd./Sci.*, 25, 585–587.

Lotka, A.J. (1934). *Théorie analytique des associations biologiques*. Hermann, Paris.

Lotka, A.J. (1956). *Elements of Mathematical Biology*. Dover Publications, New York.

Lück J. (1977). Complémentarité apico-basale dans l'organisation des tissus végétaux. Thesis, University Aix-Marseille III.

Lush, J.L. (1937). *Animal Breeding Plans*. Collegiate Press Inc., Boston.

May, R.M. (1981). *Theoretical Ecology*, 2nd edition. Blackwell, Oxford.

Mikulecky, D.C. (1996). Complexity, communication between cells, and identifying the functional components of living systems: Some observations. *Acta Biotheoretica*, 44(3/4), 179–208.

Murray, J.D. (2003). *Mathematical Biology*, 3rd edition (2 volumes). Springer, New York.

Nakielski, J., Hejnowicz, Z. (2003). The description of growth of plant organs: A continuous approach based on the growth tensor. In *Formal Description of Developing Systems*, Nation, J.B., Trofimova, I., Rand, J.D., Sulis, W. (eds). Kluwer Academic, Dordrecht.

Okabe, T. (2011). Physical phenomenology of phyllotaxis. *Journal of Theoretical Biology*, 280(1), 63–75.

Okabe, T. (2015). Biophysical optimality of the golden angle in phyllotaxis. *Scientific Reports*, 5, art. 15358.

Parrochia, D. (1993). *Philosophie des réseaux*. PUF, Paris.

Piaget, J. (ed.) (1967). *Logique et connaissance scientifique*. Gallimard, Paris.

Poincaré, H. (1968). *La Science et l'hypothèse*. Flammarion, Paris.

Ricard, J. (1999). *Biological Complexity and the Dynamics of Life Processes*. Elsevier, Amsterdam.

Ricard, J. (2003). What do we mean by biological complexity? *C.R. Acad. Sci.*, 326(2), 133–140.

Ricard, J. (2006). Statistical mechanics of organization, information, and emergence in protein networks. *Journal of Non-Equilibrium Thermodynamics*, 31, 103–152.

Ricard, J. (2008). *Pourquoi le tout est plus que la somme de ses parties : Pour une approche scientifique de l'émergence*. Hermann, Paris.

Segel, L.A. (1980). *Mathematical Models in Molecular and Cellular Biology*. Cambridge University Press, Cambridge.

Sellier, P. (1976). *Pensées de Pascal*. Mercure de France, Paris.

Soka, R., Sneath, P. (1973). *Numerical Taxonomy*. W.H. Freeman and Company, San Francisco.

Teissier, G. (1937). *Les lois quantitatives de la croissance*. Hermann, Paris.

Thieffry, D., de Jong, H. (2002). Modélisation, analyse et simulation des réseaux génétiques. *Méd./Sci.*, 18, 492–502.

Throughton, A. (1960). Further studies on the relationship between shoot and root systems of grasses. *Grass and Forage Sciences*, 15(1), 41–47.

Thom, R. (1972). *Stabilité structurelle et morphogénèse*. W.A. Benjamin, San Francisco.

Thomas, R. (1979). *Kinetic Logic*. Springer, Berlin.

Thomas, R., D'Ari, R. (1989). *Biological Feedback*. CRC Press, Boca Raton.

Thompson, D'A. (1994). *Forme et Croissance*. Le Seuil, CNRS, Paris.

Varenne, F., (2010). *Formaliser le vivant : Lois, théories, modèles ?* Hermann, Paris.

Varenne, F., Silberstein, M., Dutreuil, S., Huneman, P. (eds) (2014). *Modéliser et simuler*, vol. 2. Éditions matériologiques, Paris.

Volterra, V. (1931). *Leçons sur théorie mathématique de la lutte pour la vie*. Éditions Jacques Gabay, Sceaux.

Volterra, V. (1937). Principes de biologie mathématique. *Acta Biotheoretica*, 3, 1–36.

Zwirn, H.P. (2006). *Les Systèmes complexes. Mathématiques et Biologie*. Éditions Odile Jacob, Paris.

Index

A, B, C

allometry, *see also* biological laws, 90, 122, 123, 125–127, 144, 168
Aristotle, 2, 5, 7, 8, 10, 34, 37–39, 59, 91, 94, 136, 190
   causes, 3, 6–8, 10, 13–14, 57, 87, 95, 97, 116
   form, 5, 12, 34, 37–39, 42, 64– 67, 68–69, 83, 96, 112, 119, 137, 154, 190
   hylomorphism, 38
artificial life, 109, 110
artificial neurons, 92, 111, 112
attraction, *see also* dynamic systems (attractor), 7, 72, 119, 133, 202
automatons, 70–74, 77, 82, 89, 90, 92, 105, 109, 110, 174, 184, 192–194
   biomimetic, 109, 110
   learning, 100, 110–112
Bernard C., 1, 4, 9, 11–12, 16, 19, 30, 34, 39, 58–60, 63, 90, 95, 116, 206
   autopoiesis, 3, 59
   homeostasis, 15, 59, 104, 194, 198
bifurcation, 16, 22, 28, 43, 72, 92, 108, 148, 166, 173, 202, 204
bioinformatics, 69, 111–113

biology
   molecular, 12, 13, 52, 78–82
      genetic information, 80–84, 87, 112
      genetic program, 81, 83
      linguistic model, 83
   relational, 90, 92, 93
   systemic, *see also* integrative, 45, 78, 89, 118, 196
Buffon G.L., 17, 45–48, 64, 83, 105, 189
   Buffon's needle, 46, 47
   epistemological position, 47
   internal mold, 64, 83, 190
causality, 4–7, 10, 12, 38, 45, 47, 73, 74, 81, 86, 91, 96, 98, 112, 116, 117, 121, 197
   circular, 7, 74, 98
communication, *see also* information, 33, 73, 84– 86, 99, 110, 184
cybernetics, 7, 61, 70, 73, 74, 87, 89, 90, 109, 110, 139, 148

D, E, F

Descartes R., 6, 8, 17, 43–45
discrete versus continuous, 192
emergence, 2, 33, 59, 88, 93–96, 137, 139, 172, 193, 197, 211

environment-dependence, 10
Fibonacci sequence, 39–43, 136, 139, 174
finality, *see also* teleology, 8, 10, 11
formal genetics, 49, 50, 52, 119
   genetic algebra, 55
   genetic identity, 54
      Hardy–Weinberg's law, 51
      Mendel's laws, 50, 52, 131, 132
formal grammar(s), 74, 113, 139
game theory, 47, 105, 106, 108, 109
golden angle, 41, 136, 139

H, I, L

Heraclitus, 37, 194
hypercycle, 7, 88, 89, 171
hypothesis, 20, 27, 53, 79, 96, 101, 105, 128, `37, 139, 140, 148, 153, 154, 158, 169, 174, 190, 195, 206, 209, 211–213
information, *see also* communication, 2, 6, 12, 58, 80–87, 96, 100, 112, 166, 169, 173, 213
integrative, *see also* systemic biology, 196
invariant, 13, 28, 31, 61, 66, 101, 140, 141, 159, 168
law(s)
   biological, *see also* allometry, 15
      statistical laws, 47, 121, 128
      theoretical laws, 121, 131
   growth, 48, 98, 123, 131, 179
      Gompertz law, 49
      logistics law, 39, 49, 123, 153
Leslie matrices, 63, 196
logic kinetics of regulation, 184, 186
L-systems, 32, 72, 74, 75, 77, 174, 184, 193, 196

M, N, O, P

microscopic versus macroscopic, 195
model(s)
   probabilistic, 63
      birth-death process, 179, 182

      structured, 98, 120
   Poisson process, 181
morphogenesis, 8, 22, 29, 48, 63, 64, 67–70, 72, 74, 77, 95–97, 113, 136, 140, 144, 147, 153, 164–168, 172 174, 193, 196, 203, 209
   D'Arcy Thompson, 64–68, 136, 144
   system(s)
      transformation, 67
      Turing's reaction-diffusion, 69, 70, 175
multistationarity, 28, 31, 104, 133, 148, 166, 210, 213
networks, 7, 14, 73–77, 82, 86, 91, 92, 98–102, 104, 105, 111, 184, 185, 198, 204
notion
   of law, 36, 43
   of life, 1, 3, 8
optimality, *see also* stability, 60, 61, 97, 140, 154, 159
Pascal B., 32, 44, 198
   spirit of finesse/geometry, 198
Petri net(s), 74–78, 83, 98
phyllotaxis, 35, 41, 43, 48, 135–140
Plato, 22, 32, 37, 66, 73
position, 68, 69, 98, 137–140, 165, 166, 169, 173, 174
   constructivist, 33
   realist, 32, 178
process
   spatio-temporal, 144, 163
      electrophysiological, 175, 176
      field models, 165, 168, 175
      growth–diffusion–advection, 163
      reaction-diffusion, 69, 70, 98, 137, 140, 164, 167, 172, 173, 175
      spatialized models, 77

S, T, V

self-organization, 59, 73, 86–88, 96, 104, 105, 112

space, 26, 30, 33, 39, 133, 137, 140–147, 172, 175
　metric, 39, 143–146
　non-metric, 143, 146
spiral, 41, 42, 65, 66, 137–139
　golden, 41, 42
　logarithmic, 42, 65, 66
stability, *see also* optimality, 11, 27, 28, 31, 61, 62, 64, 69, 73, 80, 104, 133, 136, 148, 154–157, 159, 160, 162, 171, 184
　optimal control, 29, 61, 63, 159,
　paradox of enrichment, 156
　structural, 27, 31, 62, 86, 104, 154, 156, 157
statistical regression, 122

system(s), 2, 10, 16, 26, 29, 61, 69, 74–93, 104, 143, 154
　complex, 5, 6, 11, 27, 64, 79, 80, 83, 88, 93, 104, 166
　dynamic, 19, 61, 62, 98, 154, 172, 184
　　attractor, *see also* attraction, 16, 155, 157, 162
　　limit cycle, 16, 104, 108, 133, 152, 156, 157, 163
teleology, *see also* finality, 8, 10, 11
tensor calculus, *see also* tensor, 68, 215
tensor, *see also* tensor calculus, 30, 34, 68, 167, 168, 170, 171, 215
time
　biological, 142
　dependence, 82
vitalism, 8, 9, 34, 58, 95